MW01611729

SB
351
.H5
D 84
1985

Duke, James A.,
  1929-

Culinary herbs

DATE			

DISCARD

© THE BAKER & TAYLOR CO.

CULINARY HERBS: A Potpourri

CULINARY HERBS
A Potpourri

James A. Duke, Ph.D.
Chief, Germplasm Resources Laboratory, U.S.D.A.

Illustrated by
Peggy-Ann K. Duke

TRADO-MEDIC BOOKS
A Division of Conch Magazine Ltd., Publishers
Owerri   New York   London
1985

Copyright © 1985 by Conch Magazine, Limited. All rights reserved. No part of this publication may be reproduced, stored in a retrieval system, or transmitted in any form or by any means, electronic, mechanical, photocopying, recording or otherwise, without the prior permission of Conch Magazine, Limited (Publishers), 102 Normal Avenue, Buffalo, New York 14213.

First published 1985

Printed in the United States of America

Design and Composition by Conch Typesetting and Graphic Services
Dan Robertson, *designer*
Marlane Baske, *typesetter*

Library of Congress Cataloging in Publication Data

Duke, James A., 1929-
  Culinary herbs.

  Bibliography: p.
  1. Herbs. 2. Herb gardening. 3. Cookery (Herbs)
4. Herbal teas. 5. Herbals. I. Title.
SB351.H5D84   1984   635'.7   84-8684
ISBN 0-932426-32-8(U.S.)
ISBN 0-932426-33-6 (U.S. : pbk.)

WARNING: Neither the author, his sponsors nor his publishers recommend or endorse herbal medicines, self medication nor pesticides. Material relating to any of these subjects in this book is gleaned from literature such as that cited in the references. The author, a botanist, does no original research on pesticides or medicine.

CONTENTS

CHAPTER 1

An Herbalist's Dozen[1]

Beginning a book without a set of definitions is almost as risky as defining an herb. An herbal is a book in which herbs (non-woody plants), or plants in general, are named, described, and often figured, usually with special reference to their official or medicinal properties. Culinary means "of or relating to the kitchen or cookery." Emphasis, then, in this little book will fall on culinary herbs, mostly those used in cookery, or in herbal teas or liqueurs. There's not as fine a line as we'd like between the spices and the culinary herbs. The USDA once (in CA 62-24) defined spices as parts of plants, (dried seeds, buds, fruit or flower parts, bark or roots) usually of tropical origin. They contrasted herbs as from leafy parts of temperate zone

[1] Jim Duke, presented to the New Jersey Vegetable Growers Association, Atlantic City, New Jersey, January 19, 1984. With acknowledgements to Elmo Davis, McCormick Co.; Holly Shimizu, National Arboretum, USDA; and Kent Taylor, Taylor's Herb Gardens, for constructive criticisms and suggestions.

Table 1. Evaluation of Herbs[2]
Results of Twenty
Questionnaries

| Herbs | Number Times Listed | Number Times Listed First | Rank | Results at My House | |
				Jim's Ranking	Peggy's Ranking
anise	1	0	24th		
BASIL	17	1	6th	9th	4th
bay	12	0	10th		
CARAWAY	5	0	17th	10th	
catnip	4	0	19th		
CHIVES	17	1	5th	4th	2nd
coriander	5	1	16th		
corsican mint	1	0	25th		
cumin	4	0	18th		
DILL	17	0	8th	7th	6th
FENNEL	3	0	20th		10th
GARLIC	17	4	3rd	1st	1st
hot pepper	14	5	2nd		
MARJORAM	8	0	14th	6th	
OREGANO	18	1	4th	2nd	3rd
PARSLEY	19	7	1st	3rd	7th
peppermint	12	0	11th		
rosemary	11	0	12th		
saffron	2	0	22nd		
SAGE	17	0	7th	8th	8th
savory	1	0	23rd		
spearmint	6	0	15th		
TARRAGON	8	0	13th		9th
THYME	13	0	9th	5th	5th
watercress	3	0	21st		

[2]Herbalist's Dozen indicated by captial letters.

plants. Centuries ago, any plant that was not a shrub or tree was considered an herb - and suitable for experimentation to see what use might be made of it. To some botanists, a plant that dies to the ground after its yearly growing season (and, if it is a perennial, sprouts again the following spring), is an herb. To most people, an herb is a small plant, usually easy to cultivate, that offers a special usefulness. It may be grown for its aroma or culinary applications, for its medicinal virtues, or even for its historical associations. The broadest definition of herb I've found is "a useful plant." All plants are useful. Narrowing it down, an herbal use is a little known use of a well-known plant, like the diuretic use of cornsilk, or the fairly well-known use of little-known plants, like oregano for pizza pies. Perhaps King Charlemagne's definition (ca 880 AD) was best "The friend of the physician and the pride of cooks", the latter especially important in this book.

Recently, I was asked to prepare a short account of the more popular culinary herbs for the home garden in the United States. What are the most popular culinary herbs in the United States? To answer that question, I sent out a list of 50 herbs and asked respondents to check off or rank twelve easily grown herbs they considered most important to their kitchen. Their ideas did not differ radically from the list achieved at home, by taking my wife Peggy's top ten and my top ten. Her top ten plus my top ten added up to only a dozen. These dozen herbs are considered GRAS (Generally Recognized As Safe) by the FDA (Food and Drug Administration). That dozen, which I call the herbalist's dozen, is in capital letters in Table 1.

For their "Six Herbs", Avard, Story and Wentworth-Jackson (1982) sought plants "readily identifiable to the public, tolerant to water stress and low light for indoor growth, a quick germination period and fast growth rate as well as general plant hardiness." They chose half my dozen, i.e., basil, chives, marjoram, parsley, sage, and thyme. Eight (basil, dill, marjoram, oregano, parsley, sage, tarragon, and thyme) were among the 14 covered by Greenhalgh (1979) in a study of "The Market for Culinary Herbs."

To judge from my mail and phone calls, the American consumer is interested in how to grow and to use these dozen herbs. Consequently, I have drafted short writeups on their culture and use. Although I share the government's negative

3

Table 2. An Herbalist's Dozen

	Dur	Height inches	Soil	Situation	Plant	In-row Dist. inches	Between Row Dist. inches	Part Harvested	When to Harvest
1. Basil	Ann	24	Moist, fertile	No wind; sunny	Sd-Spring	6-8	18-36	Lvs, Tops	One mo. until frost
2. Caraway	Bi	30	Fertile upland	Light shade OK	Sd-Fall	8-12	24-30	Seed	Second year
3. Chives	Per	15	Variable	Light shade OK	Rt-Spr./Fall Sd-Spring	4-12	12-24	Leaves	May until frost
4. Dill	Ann	30	Well-drained	Sunny	Sd-Spring	3-12	12-24	Lvs, Seed	Lvs 6 weeks to frost (seeds when ripe)
5. Fennel	Per	36-72	Rich, well-drained	Light shade OK	Sd-Spring Pl-Spring	4-12	18-36	Lvs, Seed	Leaves all summer (seeds when dry)
6. Garlic	Per	36	Rich, damp	Sunny	Rt-Spr./Fall	3-6	12-36	Bulbs	When leaves yellow
7. Marjoram	Per	8-12	Average	Sunny	Sd-Spring	1-8	15-30	Lvs, Buds	From budding to frost
8. Oregano	Per	12-24	Poor, dry	Sunny	Sd-Spring	8-12	12-32	Lvs, Fls	Summer until frost
9. Parsley	Bi	6-12	Rich, moist	Some shade OK	Sd-Spr./Sum.	6-9	18-24	Lvs	Summer until dieback
10. Sage	Per	12-24	Poor, dry	Sunny Pl-May/June	Sd-April/May	12-24	24-36	Lvs	Summer until dieback
11. Tarragon	Per	24-36	Poor, dry or well-drained	Sunny	Sd-Spring	12-18	12-36	Lvs	Summer until dieback
12. Thyme	Per	6-12	Poor, dry	Sunny	Pl-Spring	10-24	18-36	Lvs, Fls	Summer until dieback

Column 1 = Duration, Annual, Biennal, or Perennial (at least in the deep south)
Column 2 = Usual Height of Herb
Column 3 = Soil
Column 4 = Situation
Column 5 = Planting Sd = Seed, Rt = Root, Pl = Plant
Column 6 = In-row Spacing
Column 7 = Between Row Spacing
Column 8 = Part Harvested (Lvs = Leaves, Fls = Flowers)
Column 9 = Time of Harvest

4

view of self medication and folk medicines, I admit consumers are interested in these potentially hazardous aspects of herb lore. For informative purposes only, I close each writeup with a terse listing of the reputed folk medicinal applications of the culinary herbs. Although I am aiming this treatise at a popular audience, I have used some rather unwieldy words, not wanting to discourage scholarly inquiry but hoping to discourage careless self medication. Herbalists suffering from some of the listed ailments may benefit from a placebo effect by the addition of an extra dash of the "medicinal" herb. It's surprising how many doctors will tell you that the placebo will often cure. We may even derive the placebo effect from dipping the safer herbs like snuffs. Shall we take this questionable placebo effect with a grain of herbal salt or use it in lieu of salt? It could hardly be dangerous to add an extra measure of garlic and parsley to a meal. While I can't guarantee efficacy and safety, I did not hestitate to add dill, marjoram, parsley, and thyme to the tomato juice served on New Year's Day. All are GRAS. Not suffering from a hangover myself, I still did not want to recommend anything I hadn't tried. All four are folk remedies for hangover, but hangovers, like many cases of colds and psoriasis, have no cures, at least according to modern medicine. Still, Americans spend an estimated billion dollars a year on over-the-counter "cures" for the incurable colds which affect Americans, a billion cases a year. Some quaint people would rather resort to grandmother's herbal medicine kit than throw up their hands in modern despair, admitting that there is no cure. Maybe they will find the cure in grandma's medicine chest. I certainly do not hesitate to make tea of spasmolytic herbs like basil, caraway, dill, fennel, marjoram, and thyme with anodyne herbs like fennel, when my backache flares up.

BASIL
(Ocimum basilicum L.)

CULTURE: Outside the tropics, basil grows as an *annual*, often sown from seed where it is to grow. Seed are sown (ca 5 lbs/a) about an inch apart in rows ca 2 feet apart and covered with up to ½ inch of or barely covered with finely pulverized soil. The big basil growers will want plants spaced ca 8 in apart in rows 18-36 inches apart, giving plant densities of ca 30,000-35,000 plants per acre. The small grower might alternate basil with tomato plants. They are said to make good companion crops. Seed should germinate within the first week. Some growers start their seedlings indoors for planting outside after frost (which kills the plants). Others sow many seed in 4-inch pots, getting a thick stand of seedlings to be transplanted outside. Basil roots readily from cuttings if kept moist. It grows well in moderately fertile soil, moist but well-drained, protected from the wind, in full sun or partial shade. Best quality basil is grown in full sunlight. Fertilization at ca 120:100:100 lb/a has been suggested. Plants should be thinned to 8-16 inches apart in the rows, and weeded. Like many minor crops, basil is an orphan when it comes to herbicide recommendations. In general, the mint genus, Mentha, and *perhaps* other members of the family, especially perennials, are rather "tough." Peppermint and spearmint are treated preemergence with "Goal 2E" (oxyflurofen)[3] or "Sinbar 80W" (terbacil) and postemergence with "Basagran" (bentazon). "Treflan 4EC" (trifluralin) has been used on dormant peppermint and spearmint. Growers could carefully experiment with small plots to determine what will work best in their situation. Plants may need little more attention until harvest time. Leaf yields range from 1-3 tons/acre, dried down from 6-10 tons/acre of fresh herb.

---

[3] Mention of or failure to mention a given product does not constitute endorsement or disendorsement of any product mentioned in this book, by the author, his publisher, or his sponsors. Herbalists, hobbyists, and organic gardeners will probably prefer manual weeding to herbicides. Personally I recommend hand weeding for any crop I intend to ingest. No one should use any pesticide without following all the instructions carefully. For many herbs there are no legally approved pesticides.

BASIL

7

Tops of the branches should be pinched off to deter flowering and to induce bushier branching habit. The pinchings can be used herbally like your final leaf harvest, fresh, air-dried, sun-cured, or frozen. The home grower can harvest basil leaves and flowering tops right up to frost. Any time you have a super-abundance of basil, you can cut off the tops, tie several together by their stalks, and hang them in an airy place out of the sun, for use throughout the winter. Such an approach will work for any of the leafy herbs. Alternatively, the leaves can be stripped off the growing plant in the field, or back in the drying room. American producers dry the leaves with artificial heat not exceeding 40°C. Sun-dried leaves tend to become brownish. You can prepare basil (or other herbal) salts as you dehydrate your basil (or other herbs). Sprinkle some table salt on a cookie-tin, then flatten out fresh leaves on the salt, sprinkle on another layer of salt, and place in an oven for 10 to 20 minutes at about 300°F. Then remove the large pieces of leaves and crush by hand or with a mortar and pestle. This constitutes your powdered or flaked herb. The salt constitutes herb salt. If you should somehow scorch your leaves, don't throw them out without first experimenting with them. Early Americans used certain charred leaves as substitutes for ''instant'' coffee, or ashes for salt or soap. Charred or ashed herbs might be so used. Basil and other sweet herbs can also be dried with sugar, generating herb sugars as byproducts. Fresh basil tops and/or leaves may also be put in polyethylene bags and frozen for winter use, without blanching. Extra basil might be added to vinegar for preservation. Italians preserve their fresh basil in salad oil. Purple or opal basil adds nice pink color to vinegars in which it is steeped, creating an interesting basil vinegar.

There are several variants of basil. *Ocimum basilicum* 'Minimum', a smaller version, makes an excellent pot plant and should be grown where space is limiting. Lemon basil, *Ocimum basilicum* 'Citriodorum', combining the fragrance of lemon and spicy basil, is excellent in cooking. Other forms include 'Dark Opal', 'Spice Basil', and 'Cinnamon Basil'.

USES: Basil salt is excellent on tomatoes, basil vinegar on lettuce and tomato salad, basil oil on tossed salads. These uses, of course, reflect basil's major culinary role as a companion of tomato. Basil is also used to enhance the flavors of many vegetables: artichoke, asparagus, beans, beets, broccoli, Brussels sprouts, cabbage, carrots, cauliflower, celery, cucumber, eggplant, mushrooms, peas, peppers, potatoes, spinach, tomatoes, turnips, and zucchini; and meat (beef, brains, chicken, eel, fish, ham, lamb, liver, mullet, partridge, pheasant, pork, rabbit, sausage, shrimp, turkey, turtle, veal, and venison). For those limiting pepper in their diet, basil serves as a substitute. Dishes like minestrone, pasta, pizza, and spaghetti are also improved with basil. Dried basil leaves are used to flavor dressings, omelets, rice, soups, and stews. The essential oil of basil finds its way into cordials, cosmetics, perfumes, soaps, and spices. Basil is also used in herb teas and liquers. "Bloody Marys" are said to be improved with basil and parsley. Basil hung around the neck is said to repel flies.

FOLK MEDICINE: Reported to be antispasmodic, alexeteric, anodyne, aphrodisiac, carminative, cyanogenetic, demulcent, diaphoretic, digestive, diuretic, expectorant, lactagogue, laxative, pectoral, refrigerant, sedative, stimulant, stomachic, sudorific, and vermifuge, basil is a folk remedy for alcoholism, anasarca, boredom, cancer, catarrh, cephalalgia, cholera, colic, collapse, constipation, convulsion, cough, croup, deafness, delirium, depression, diarrhea, dropsy, dysentery, earache, enteritis, epilepsy, fever, flu, frigidity, gastroenteritis, gonorrhea, gout, gravel, halitosis, headache, hemiplegia, hiccup, hysteria, impotency, infection, inflammation, insanity, insect bites, labor, migrain, nausea, nephrosis, nerves, paralysis, parturition, piles, polyps, ringworm, snakebite, sinusitis, sores, sore throat, spasm, stings, stomach, throat, toothache, tumors, urinary ailments, wart, whooping cough, and worms (Duke and Wain, 1981). Basil infusion is taken for halitosis, headache, and gout. The leaf juice may allay throat irritation.

CARAWAY
(Carum carvi L.)

CULTURE: A *biennial*, flowering and dying in its second season, caraway does well on well-prepared, upland fertile soil. Seeds may be sown (ca 4-6 lbs/a) in spring or autumn in rows, and thinned to 8 to 12 inches apart. There should be 2-3 plants per foot in rows about 30 inches apart. Since seed will not mature until the second season, caraway is an ideal candidate for intercropping in the home garden with such annuals as beans, coriander, dill, mustard, or peas. After the intercrops are harvested the first year, the caraway need only be kept free of weeds until it matures the second year.

Large growers are mechanized; from mechanical drilling of the seed, mechanized cultivation, on to mechanized harvesting and threshing. The small herbalist may cut his plants to dry, using a sickle or scythe, or may pull up the plants, or cut off the tops, one by one. Seed yields usually run about 500-2000 lbs/acre. There is a residual ton of straw, which serves as animal food (Rosengarten, 1973).

Seed harvested in summer of the second season must be dry and stored in a dry place. Ripe seed can be dried in the sun or over a low heat, stirring occasionally. Some herbalists recommend that the seeds by sterilized with scalding hot water to rid them of stored seed insect pest. Seed can also be saltcured, frozen, or steeped in vinegar.

USES: Most familiar to Americans as the seed on rye bread, caraway has a host of culinary uses. Caraway is traditional with beet, cabbage, carrot, and is often served with baked apples, other fruits, and breadstuffs. It is used with goulash and roast pork. Cheeses, soups, and vinegars are also flavored with caraway. Several liquers utilize caraway, among them Aquavit, Danzig, Danzigwasser, Goldvasser, Kummel, and L'huile de Venus, and certain types of Schnapps. In Scotland, even today, buttered bread is dipped into a saucer of caraway seed. In the Middle Ages, and elsewhere today, caraway leaves were chopped up and used instead of parsley.

10

CARAWAY

FOLK MEDICINE: Regarded in folk medicine as anti-spasmodic, bactericidal, carminative, diaphoretic, digestive, diuretic, emmenagogue, expectorant, fungicidal, lactagogue, laxative, stimulant, and stomachic, caraway has been used to alleviate bruises, cancer, cholera, colic, dysmenorrhea, dyspepsia, earache, fistula, flatulence, halitosis, headache, hookworm, hysteria, impotency, incontinence, indigestion, nausea, prolapse, scabies, sores, spasms, stomach ailments and syphilis. Caraway oil has antihistaminic, antispasmodic, bactericidal, fungicidal, and larvicidal properties.

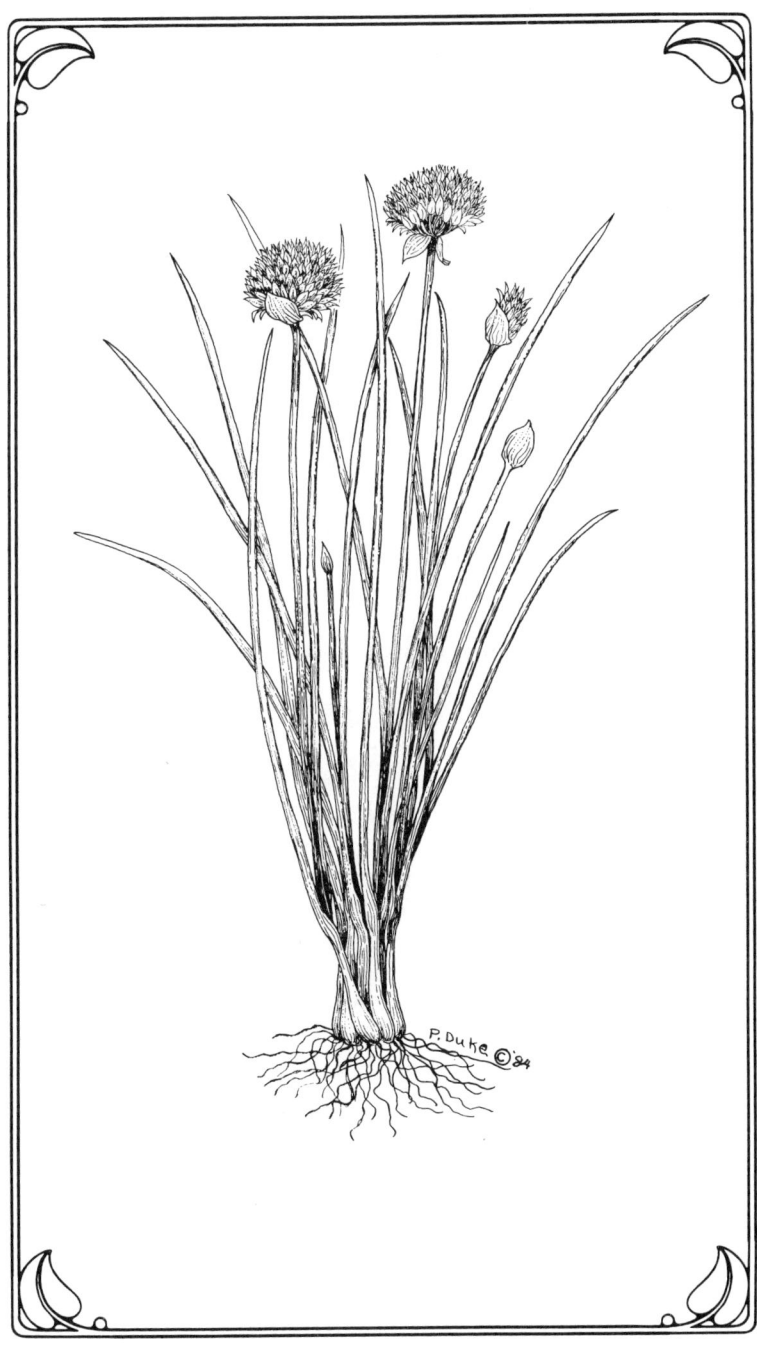

CHIVES

CHIVES
(Allium schoenoprasum L.)

CULTURE: As *perennials*, chives need a fertile well-prepared, well-weeded bed with full sunlight, but will tolerate most garden soils. They can be started from seed but plants so started should not be harvested the first year. The following year, the one-year-old plants can be divided into smaller clumps for new plants, and spaced 4-15 inches apart in rows ca 20 inches apart, allowing for careful weeding. Densities of 40,000-60,000 plants per acre are recommended. Second year plants may be clipped for use in the kitchen. It is better to cut a few side leaves down to ground level than to clip all the leaves halfway. Using the former approach, the plants can be clipped again and again and continue to thrive. Such plants should be fertilized with commercial fertilizers, bone meal, or sterile manures. Beds should be dug up, divided, and replanted every two to four years. With repeated clippings, chives, especially Chinese chives, can yield up to 4 tons per acre.

USES: Harvested leaves, cut up fresh in salads, soups (e.g., asparagus, beans, cauliflower, potato, tomato, vegetable), and stews, may be added to vegetables in melted butter, or kneaded into cheeses or doughs. Chives are added fresh to such vegetables as asparagus, beets, carrots, corn, mushrooms, peas, potatoes, and such salads as lettuce and tomato, cucumber salad, egg salad, endive salad, potato salad, slaws, tomato and cheese salad, and such meats as bacon, beef, chicken, duck, fish, ham, pork, sausage, steak, etc. Many dishes benefit from the judicious use of chives. Hot salted broths using several of the nonsweet herbs like chives, dill, garlic, oregano, and sage, make an interesting pickup on a cold day. Biscuits, canapes, cheeses, croquettes, fritters, omelets, and vinegars all can be improved with chives. Chive flowers, added to vinegar, yield a pinkish or lavender color.

FOLK MEDICINE: Chives figure in folk remedies for anemia, bad appetite, blood disorders, hypertension, kidney, and stomach ailments, and worms. Perhaps because they are so much milder than garlic and onion, which belong to the same genus, Allium, chives have not been so widely heralded in folk medicine. Quite possibly, the chemistry and folk applications of thes three aromatic species are similar in many instances. Chive roots, after cooking in water, are said to alleviate coughs.

DILL
(Anethum graveolens L.)

CULTURE: The large-scale grower will find it convenient to drill seeds of this *annual* in rows 1 to 3 feet apart, in late fall or early spring 5-10 lbs/a. The small gardener should cover his seed very shallowly with well-prepared sandy soil in well-drained situations with full sun. Dill is said to exhaust the soil. Seed will sprout within the week. Growing quickly in the cool weather, maturing in as few as 6 weeks, once established in rows 12-24 inches apart, the plants are thinned to 3-12 inches apart. Plant densities of ca 100,000 plants are recommended. Dill is said to be a good companion plant for cabbage. Dill might be successfully intercropped with grapes, since dill controls or alleviates some grape mildews. Dacthal, Sonalan, and Treflan have given satisfactory weed control (Precheur and Garrabrants, 1983). It can be interspersed with row crops, as long as members of its own family (caraway, cumin, carrot, fennel, etc.) are avoided. If dill is pinched back to prevent flowering, it can sometimes be treated as a biennial. The herbage pinched back could be as useful as the mature dillweed. Once hot weather has set in, pinching back will not stop flowering. After it flowers and seeds, dill dies. Extreme heat, hail, heavy rain, and strong winds can ruin the crop. Seed yields run 500-1000 lbs/acre.

Commerical farmers may seek dill herb oil or dill seed oil, while the gardener may seek the herbage and seeds. Herbage can be harvested and dried starting about 5 weeks after germination, but it may take 4 months for the seed to mature. Leaves may be processed like basil, seeds like caraway. The essential oil is both repellent and toxic to the granary weevil.

USES: Best known for its role in pickle processing, dill has a host of culinary attributes. Dill is recommended for diabetics and those on saltfree diets, (herbs can improve saltfree dishes your doctor may have recommended). Dill has been added to such fruits and vegetables as artichoke, asparagus, avocado, beans, beets, broccoli, Brussels sprouts, cabbage, salsify, spinach, tomato, turnips, and watercress; and such salads as cabbage salad, cucumber salad, fish salad, and potato salad;

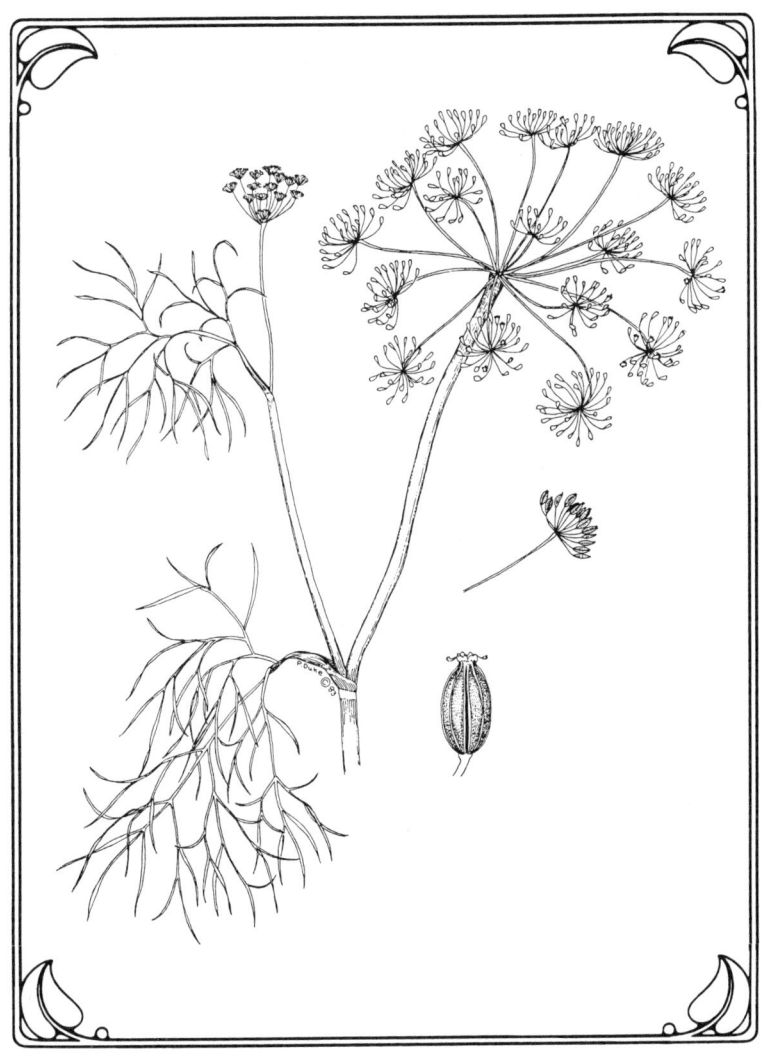

DILL

17

and bean, chicken, fish, pea, potato, tomato, and vegetable soups; meats like chicken, corned beef, crayfish, eel, fish, ham, lamb, liver, lobster, pork, shrimp, snails, and even steak benefit from dill, alone or in combination with other herbs. Dillseed are good with cabbage, cucumber, or potato salads, and are chewed to improve the breath; herbage and seeds are good in butters, cheeses, and dips. Wilder recommends sprinkling the seed on green apple pie. Roasted seed have served as a coffee substitute.

FOLK MEDICINE: The seed and/or leaves, prepared in various manners, are used in folk remedies for abdominal tumors, condylomata, indurations, and tumors (abdomen, anus, liver, mouth, stomach, throat). The flower, cooked in oil, is said to alleviate anal tumors. Dill is considered aperient, balsamic, carminative, detersive, digestive, diuretic, lactagogue, laxative, narcotic, psychedelic, resolvent, sedative, stimulant, stomachic, and tonic. Dill water is used for children's ailments, like flatulence and indigestion. The plant figures also into folk remedies for anemia, bruises, colic, cough, cramps, dropsy, halitosis, hangover, hemorrhoids, hiccups, indigestion, insomnia, jaundice, sclerosis, scurvy, sores, and stomachache. Dillseed oil is spasmolytic and bactericidal.

FENNEL

FENNEL
(Foeniculum vulgare Mill.)

CULTURE: Easily started from seed or root divisions, this hardy *perennial* (often treated as annual or biennial) can grow to 6 feet tall with little or no attention after the first weeding, surviving partial to half shade. Seeds may be broadcast or drilled at 2 to 10 lb/a. Commercial growers drill seed in rows 18 to 36 inches apart, thinning the seedlings to 4 to 12 inches apart in the row. Densities of 40,000-60,000 plants per acre are suggested for fennel seed. Weeding and irrigation during drought are recommended. Tolerating poor soil and a smoky atmosphere, as well as partial shade, fennel is fairly well adapted to window gardens. Well drained loams or black sandy soils with sufficient lime are recommended. Those interested in herbage should pinch off the flowers, while those interested in seed should get their plants off to a dangerously early start in the spring. Fully ripe yet green seed should be slowly dried in the dark or shade, preventing excess moisture from accumulating. The herbage can be used fresh or processed like basil. Seed yields may exceed 1000 lbs/acre.

USES: Probably best known as a good herb for cooking with fish, fennel is a very popular culinary item, useful in salads, soups, stews, and as an adjunct to fish and meat dishes, as well as countless vegetables, (e.g. asparagus, beans, beets, cabbage, carrot, cauliflower, eggplant, kraut, lentils, mushrooms, peas, potatoes, squash, tomatoes). Beef, chicken, fish, lamb, pork, sausage, shrimp, snail, and veal recipes often call for fennel. Roman bakers are said to have placed fennel under the loaves for the flavor it imparted. Both foliage and seeds are used in tisanes, wines, and liqueurs. A pound of leaves, 4 lbs apple, 5 lbs sugar, and 1 oz citric acid are the main ingredients in 2 gallons water for fennel wine. Aerobic fermentation for 3 to 5 days should do the trick. The liqueur *fenouillette* is based on fennel. Fennel figures in some types of absinthe, celery, and kummel liqueurs. La Tintaine is a French liqueur flavored with anise and fennel, served in a treeshaped bottle. I prefer fennel to anise in various recipes for food and drink because I've had more luck growing fennel than anise. I get a hint of caramel in some fennel. Since fennel was said to impart longevity, strength, and courage, I have no objection to its inclusion in anything I eat or drink. Powdered fennel is said to repel fleas from kennels and stables. Facial packs of fennel infusion and honey are recommended for facial wrinkles. Fennel tea in yogurt is said to alleviate oily skin, possibly preventing wrinkles.

FOLK MEDICINE: Regarded as anodyne, antispasmodic, aperient, aphrodisiac, aromatic, cardiotonic, carminative, diaphoretic, digestive, diuretic, ecbolic, emmenagogue, expectorant, galactagogue, pectoral, stimulant, stomachic, tonic, and vermicidal, fennel has been recommended for aerophagia, amenorrhea, anemia, asthma, backache, bronchitis, cholera, colic, coughs, cramps, ears, enteritis, enuresis, eyestrain, flatulence, gall bladder, gas, gastritis, halitosis, headache, hoarseness, hookworm, indigestion, insect bits, intestinal and stomach ailments, kidneys, liver, measles, migraine, nephrosis, nerves, obesity, parturition, rheumatism, sinusitis, smallpox, snakebite, sores, spasms, splenosis, stings, stomach, stones, strangury, tenesmus, toothache, etc. The oil contains antioxidants.

GARLIC
(Allium sativum L.)

CULTURE: Rich well-drained organic soils are appropriate for this *perennial*. Garlic cloves are planted 1 to 2 inches deep, 3 to 6 inches apart in rows 1 to 3 feet apart. Densities of 100,000-150,000 plants per acre are suggested. Spring planting is recommended in the north, fall planting in the south. To keep weeds down, 4-10 lbs/a DCPA (dacthal) is recommended over transplants at planting time or at the 3-5-leaf stage. By the time the foliage has died down in the summer, each clove will have multiplied to form several cloves, each of which can be used for the kitchen or for replanting. Plants should be kept moist early in the season, but irrigation should be discontinued as the plants mature. Organic fertilizer and/or a 5:10:5 mineral fertilizer (ca 1000 lb/a) are recommended.

It may take 4 to 8 months for garlic cloves to mature. They should be harvested after the leaves start dying back, carefully pulled during a dry spell and left to dry in the field, or moved to well-aerated racks. They can be hung and, like other herbs, must be kept dry, or they may mildew. U.S. production figures at about 5000 pounds per acre, but 5 tons or more is not unusual.

USES: Excepting garlic's brother the onion, usually considered vegetable rather than herb, no other herb assumes such culinary importance. Garlic is indispensible to Latin cuisine. For use, the garlic may be diced, or merely rubbed onto the container in which a dish is being prepared. It mixes well with some vegetables, salads, and soups (e.g. bean soup, broccoli, carrots, dandelions, eggplants, endive, mushrooms, potato salad, squash, zucchini), and many meats (beef, fish, lamb, mutton, pork, scallops, shrimp, etc.). A garlic butter to coat a lobster or just to spread on bread is an inexpensive but elegant addition to any table. Much as this American considers as null and void a beef stew without both cooked and raw onions, many Europeans relish fish chowder with cooked garlic within, raw garlic added, and garlic-coated croutons. Cooked and raw garlic are very different from a culinary point of view. For those whose spouses or lovers object to their garlic breath, celery and

GARLIC

23

parsley are recommended. Italians feed garlic to their pigs as a growth-stimulating antibiotic in lieu of zinc bacitracin.

FOLK MEDICINE: Few herbs have been assigned so many medicinal attributes, and there is a growing body of evidence to suggest that some of these folk features have scientific rationales. Supposed to lower the blood pressure, garlic might be recommended to those millions of Americans suffering hypertension. Garlic vinegar is said to have imparted an immunity to plague in Europe. Made less offensive with caraway and fennel seed, and sweetened with honey, garlic vinegar becomes the garlic syrup, once recommended for asthma, bronchitis, coughs, and difficult breathing. Garlic is considered antioxidant, antiseptic, carminative, cholagogue, depurative, diaphoretic, digestive, diuretic, emollient, expectorant, pectoral, rubefacient, stimulant, stomachic, and vermifuge. Garlic has been recommended also for such diverse ailments as angina, anthrax, arteriosclerosis, asthma, baldness, bilious ailments, bladder ailment, burns, candidiasis, colds, colic, coughs, corns, cramps, diabetes, diarrhea, diptheria, dropsy, dysentery, dyspepsia, earache, eczema, epilepsy, fever, flatulence, flu, gall bladder, gangrene, gastroenteritis, headache, heart, hysteria, impotence, indigestion, itch, liver ailments, malaria, melancholy, prostate, rabies, ringworm, rheumatism, sciatica, scrofula, sinusitis, smallpox, snakebite, stings, stomachache, thrush, tuberculosis, typhoid, ulcers (internal), urinary ailments, whooping cough, worms, and probably dozens more ailments. The more I read, the more I think garlic may play a big role in fighting *Candida,* a fungal enemy of man, perhaps one of our more mortal enemies.

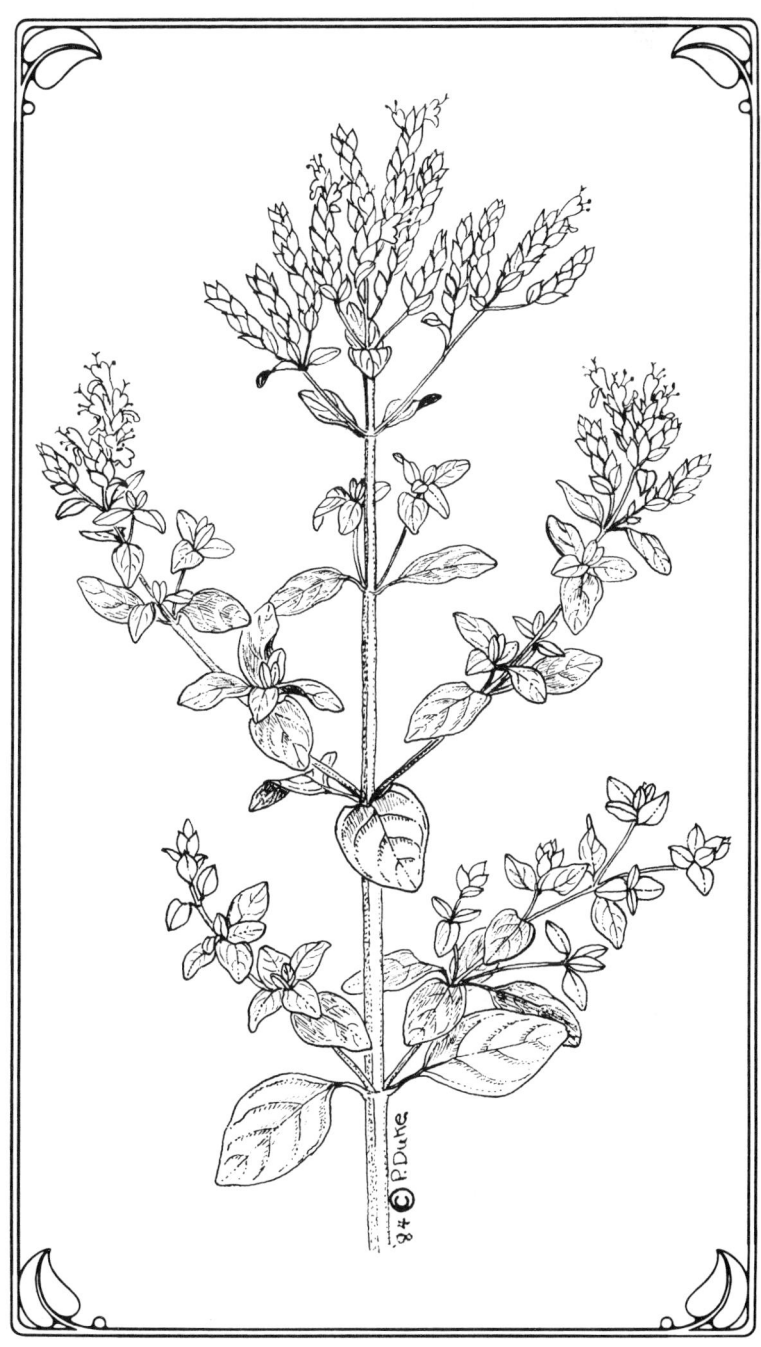

MARJORAM

25

MARJORAM
(Origanum majorana L.)

CULTURE: Tolerating some shade and a variety of soils, this *perennial* can be direct seeded into well-prepared soils and covered very shallowly. Soil must be warm and moist. Plants can be started under glass, seed germinating in 7 to 14 days, and set out at the 4- to 6-leaf stage. Plants are spaced 1 to 8 inches apart in rows 15 to 30 inches apart. Densities 15,000-40,000 plants per acre are recommended. Seedlings are very susceptible to drought, frost, heavy rain, and wind. Once established, the plants become more tolerant of these factors, and may spread to become weedy. Cuttings and subdivisions of this mint, like most mints, are readily made. Cuttings have given higher yields in marjoram, while seeding lemonbalm, sage, and thyme gave higher yields. Farther north, frost will kill back the marjoram, but down south, marjoram is **perennial**. It makes a useful intercrop in the rows of tomatoes and peppers, etc., perhaps interrupting the train of thought of vegetable insects which might mow down a monocultured row. Fertilization of ca 20-80 lbs N. 35-70 lbs P_2O_5, and 80-280 lbs K_2O have been recommended.

Once established, leaves can be harvested from preflowering plant material until frost kills the plant back to the roots. Leaves may be processed like basil. Yields of one ton or more per acre are not unexpected. Harvested leaves retain their flavor well when dried.

USES: Containing antiseptic ingredients, like many mint species, marjoram is a useful culinary herb (much sweeter than oregano), going very well with several vegetables (e.g. beans, cabbage, carrot, cauliflower, cucumber, eggplant, mushrooms, peas, potatoes, pumpkin, salsify, spinach, squash, tomato, and zucchini) and meats (beef, chicken, clam, duck, fish, goose, ham, lamb, liver, oysters, pork, steak, turkey, and veal) as well as cheeses, macaroni, omelets, pastries, salads, soups, and vinegars. Marjoram can be a bland substitute for pepper, if you're trying to do without pepper; likewise salt. Volatile compounds from marjoram extracts reduce spore germination and respiration of the fungus Fusarium. Hence the plant might be suggested as an intercrop or mulch for other crops where Fusarium is a problem. Daisley (1982) recommends the flowers and leaves as ingredients in "effective" moth bags.

FOLK MEDICINE: Regarded in folk lore as aphrodisiac, astringent, carminative, diaphoretic, diuretic, emmenagogue, expectorant, lactagogue, rubefacient, sedative, stimulant, stomachic, sudorific, and tonic, marjoram is used in folk remedies for amenorrhea, asthma, bedwetting, bronchitis, bruises, cancer, coughs, cramps, diarrhea, dysmenorrhea, dyspepsia, erotomania, fever, gastroenteritis, hangover, headache, hysteria, indigestion, insomnia, measles, nausea, nerves, paralysis, rheumatism, spasms, sprains, stomach, and toothache. Avard et al (1982) add, "It has a reputation for increasing white blood corpuscles and improving circulation."

OREGANO
(Origanum vulgare L.)

CULTURE: Tolerating most garden soils, oregano or wild marjoram, can be direct seeded into well-prepared garden soils. It is a hardy *perennial* with very small seed which can be strewn right on the surface, and stirred or firmed into the row, barely covered with finely pulverized soil. It is easily propagated by cutting and by root divisions in fall. Usually spaced 8-12 inches apart in rows 12-32 inches apart. Densities of 16,000-32,000 plants per acre are recommended. Large plants can be interspersed between vegetables in their rows, breaking up the monoculture. Yields of a ton or more are possible.

Once plants are established, leaves can be stripped off the plants in place, or the stems can be severed near the soil level, bundled up and strung up to dry in an airy dry place out of direct sunlight. The stems can be cut back, leaves stripped off, and the nude stems repotted with rooting hormone. Once a superior clone is located, clonal reproduction is recommended. Many mint species can be rooted after such rough treatment. Replanting is recommended after four or five years. Oregano might be considered more a smell than a plant. Sources of so-called "oregano" include species of Coleus, Hyptis, Lantana, Lippia, as well as many different species of Origanum. There are forms of *Origanum vulgare* which do have pleasant flavor and are useful in cooking, although the more desirable is *Origanum heracleoticum*, sometimes called Greek or Italian oregano. Extremely hardy and easy to grow, *Origanum heracleoticum* has a strong, peppery aromatic flavor. True Greek oregano is *Origanum onites*. This species is found on hillsides in Greece where it forms a subshrub. In cool climates, *Origanum onites* is not very hardy. One of the major problems with oregano in the trade is that suppliers of herbs often sell just the common oregano, *Origanum vulgare,* which has little flavor. Its only real merit is an attractive purple flower and vigorous growth habit!

USES: Attaining fame in the United States only in the twentieth century as **the** pizza herb, oregano has a wide array of culinary applications. One must remember that genera in at

OREGANO

least three families and many species are masquerading under the name "oregano". The Mexican type is used in chile con carne, chile powder, enchiladas, and tamales. It graces many vegetable recipes (asparagus, beans [dry and green], broccoli, carrots, celery, eggplant, mushrooms, peas, peppers, spinach, tomato, and zucchini), soups (bean, onion, tomato), and meats (beef, chicken, fish, lamb, liverwurst, pheasant, pork, shrimp, turkey, veal, venison). Some chefs suggest that oregano may be substituted for thyme or vice versa in culinary recipes.

FOLK MEDICINE: Regarded in folk medicine as anodyne, carminative, decongestant, diaphoretic, emmenagogue, emollient, expectorant, nervine, pectoral, rubefacient, sedative, stupefacient, sudorific, and tonic, oregano is used in home remedies for alopecia, asthma, bronchitis, catarrh, cancer, childbirth, colds, colic, coronary conditions, coughs, diarrhea, ears, fever, flatulence, flu, gastroenteritis, headache, hysteria, indigestion, insomnia, itch, jaundice, lungs, measles, nerves, rheumatism, toothache, tumors, and whooping cough.

PARSLEY

31

PARSLEY
(Petroselinum crispum Mill.)

CULTURE: Parsley, a *biennial*, does well in well-drained organic loams. Since it transplants poorly and germinates slowly, parsley is seeded (5-10 lbs/a), often mixed with radish seed, about ¼ inch deep. The radishes can be harvested as the parsley is thinned in the row. Plants are usually spaced 6-9 inches apart in rows ca 18-24 inches apart. Densities of ca 40,000 plants per acre are recommended. Too much sun or drought can harm germinating seedlings. Crowding parsley results in precocious flowering, rendering the foliage almost useless. Soil around the plants should be well-aerated and free of weeds. Cultivate only if necessary to control weeds. For weed control, a post-emergent application of stoddard solvent (40-60 gal/a) has been recommended. A 5:10:5 fertilizer, at a rate of ca 100 lb/a is recommended. Parsley plants can be kept in the kitchen window, avoiding too much heat or cold, and washing with soap and tap water to keep insects down.

Leaves can be harvested throughout the summer and dried in a dark dry place. They should be dried at closer to 200 °F, rather than the 100 °F recommended for most other foliage herbs. They can be processed like basil, allowing more time or temperature for drying. Parsley also freezes well. Parsley leaf yields should exceed a ton per acre. Seed yields are close to 1,500 pounds. Two distinct types of parsley are used. (1) The curled leaf type, important as a garnish *(Petroselinum crispum)*. (2) The flat leaf parsley *(Petroselinum cripsum* var. *neapolitanum)*, known as Italian parsley, and preferred for cooking and chopping.

USES: Few herbs are as widely used in the U.S. as parsley, except perhaps garlic. Parsley is a healthful garnish, capable of masking foul odors. The leaves are chewed to correct halitosis and to mask the smell of alcohol. Leaves and roots, fresh or fried, serve as vegetable or condiment. It is used with many vegetables (asparagus, beans, broadbeans, broccoli, cabbage, carrots, cauliflower, celeriac, corn, dandelion, endive, leeks, lentils, lettuce, mushrooms, onion, peas, potatoes, squash, tomato, and zucchini), and meats (chicken, clam, dove, fish,

32

lamb, meatloaf, rabbit, shrimp, steak, tuna, turkey, and veal). In European cookery, parsley enters egg, fish, fowl, meat, shellfish, and soup dishes. Fox (1933) says parsley soup is delicious. Parsley is also used in vegetable tisanes. In France, a mixture of parsley and shallot, finely chopped, is added as *persillade* toward the end of cooking a dish. It is important in boquet garni, in butters and vinegars, and in *ravigote, sauces, tartare, vinaigrette,* and *verte.* Mixed with bulgur wheat, it is an important middle eastern salad ingredient. It is said to be rendered more piquant by the addition of lemon balm.

FOLK MEDICINE: Parsley, pounded with snails, was applied to scofulous swellings as an ointment. Bruised leaves are used like those of celandine, clover, comfrey, and violet, to "dispel cancerous tumors" (Grieve's Herbal). It has also been used in uterine disorders. A strong root decoction is said to help kidney congestion, dropsy, gravel, jaundice, and stone. Parsley tea was once served the troops in the trenches suffering from dysentery. It has been recommended for gallstones. Bruised leaves are used to alleviate insect bites and to get rid of lice and skin parasites.

The apiol in parsley, used for ague, nervous ailments, and formerly official in the U.S. as an antipyretic and emmenagogue, can be poisonous under certain extreme conditions. In large doses, the oleoresin of parsley (apiol, apiolin, and myristicin) produces giddiness and deafness, fall of blood pressure, and some slowing of the pulse and paralysis, followed by fatty degeneration of the liver and kidney, similar to that caused by myristicin. Regarded in folk remedies as abortifacient, antibiotic, aperient, aphrodisiac, carminative, diaphoretic, diuretic, ecbolic, emmenagogue, stimulant, and sudorific, parsley is used for amenorrhea, amnesia, anemia, arteriosclerosis, asthma, bites, bleeding, boils, bruises, cachexia, cancer, conjunctivitis, coughs, dropsy, dysentery, dysmenorrhea, dyspnea, eyes, fever, flatulence, gallstones, gangrene, gravel, halitosis, hangover, heart, hypertension, indigestion, insect bites, jaundice, kidneys, liver, malaria, nausea, prostate, sores, splenitis, stings, stones, tumors, and uteritis.

SAGE
(Salvia officinalis L.)

CULTURE: Faring best on well-drained, rich clay loams, but tolerating most garden soils, sage can be started from seed or cuttings. As with other mint species, there are many chemovars (true-breeding chemical "varieties"). Once you have a good clone, clonal multiplication may be desirable. Grown as a *perennial*, plants are spaced 1 to 2 feet apart in rows 2 to 3 feet apart. Plant densities of 12,000-16,000 per acre are suggested. Weeds must be minimized, and a 2:10:10 or 5:8:7 fertilizer is recommended. Oat straw mulches have increased yields by ca 33% (1500 lbs to 2000 lbs air-cured sage per acre). Rather resistant to insects and adverse weather and drought, sage requires little attention once started. Farther north, frosts may kill this perennial.

Sage is harvested on clear dry days, cut by hand or machine, depending on the size of the operation. The small gardener can cut his plants individually near the soil line and tie them into shocks for drying out of the sun. Artificial heat is needed for drying during damp weather. Yields run from ca 250 to nearly 4000 lbs air-dried sage per acre. The tricolor variety is recommended for the kitchen window garden.

USES: Although best known as a sausage ingredient, dried sage leaves are also used for flavoring, stuffings, soups, and some canned vegetables. Sage is used perhaps more with meats (bacon, clam, crab, dove, duck, eel, fish, goose, grouse, ham, lamb, liver, mutton, pork, rabbit, sausage, tripe, turkey, veal, and venison) than with vegetables (beans, broad beans, eggplant, limas, onion, peas, squash, spinach, and tomato). Said to repress fish odors, it seems natural to add to strong fish dishes. Fresh leaves are also used in herb butters, cheeses, liqueurs, pickles, salads, and vinegars. Rabbits fed on sage develop a particular tasty meat. Chinese were said to prefer sage tea to their own true tea. An ale, made of sage, betony, scabious, spikenard, squinnette, and fennel seed was said to be quite wholesome. Sage, speedwell, and wood betony is said to make an excellent breakfast tea. Italians nibble on sage to preserve the health. Other Europeans use the leaves as a crude

SAGE

but effective toothbrush, said to remove plaque, clear stains, and stimulate the gums when rubbed gently on the teeth (Daisley, 1982). Sage butter and bread is said to be a satisfying meal in itself. Various combinations of sage, onion, butter, and lemon make excellent relishes for meat dishes. On a winter day in Holland, you might encounter sage leaves steeped in hot milk. In Greece, they'd more likely be steeped in water.

FOLK MEDICINE: Regarded in folk medicine as astringent, lactagogue, laxative, stomachic, tonic, and vermifuge, sage is used in home remedies for asthma, bronchitis, catarrh, chilblains, cold, consumption, diarrhea, dizziness, dysmenorrhea, eczema, erotomania, falling hair, fever, flatulence, flu, gallbladder, gastroenteritis, gingivitis, headache, heart, hysteria, indigestion, infection, kidneys, larynx, leucorrhea, liver, lungs, nausea, nerves, nightsweats, piles, rheumatism, quinsy, snakebite, sores, sore throat, stomach, tonsilitis, toothache, tumors, ulcers (internal), vertigo, warts, whooping cough, worms, and wounds.

A tea or liqueur made from sage, mint and elderberry flowers makes a useful antiseptic gargle for sore throat. Horehound and fenugreek are said to make useful additives to this gargle. Phenolic acids from sage are effective against Staphylococcus.

Fresh leaves make a good dentifrice. A sage gargle is recommended for bleeding gums, sore throat, and tonsilitis. The tonic tea has been recommended for kidney, lethargy, liver, lungs, measles, nerves, and phthisis. The infusion, applied to the scalp, is said to darken the hair. The oil is applied in cases of rheumatic pain.

An oil distilled from the plant is said to be a violent epileptiform convulsant, resembling the essential oils of absinth, nutmeg, and wormwood. The dried leaves have been smoked to treat asthma. Smelled for some time, sage is said to cause intoxication and giddiness. Sage extracts are said to be antioxidant due to the presence of labiatic and carnosic acids. Sage oil has neurotropic antispasmodic activities.

TARRAGON

TARRAGON
(Artemisia dracunculus L.)

CULTURE: True tarragon, a *perennial*, cannot be started from seed since it sets no seed. It is best propagated by root division in spring. It should be lifted and divided every 3 to 5 years. Early in the year, cuttings take readily with rooting hormones in well-washed sand. As the plants approach their fall dormancy, they are difficult to root. They fare best on rather dry, well-drained soils of relatively low fertility. Tolerating some shade, they do best in full sun. Home gardeners need only a few plants to intersperse among ornamentals or vegetables. Bigger growers space their plants about 1 ½ feet apart in rows about 3 feet apart. Others suggest spacings of 12-16 inches apart in rows 12-20 inches apart. Densities of 25,000 to 30,000 plants per acre are recommended. Once started, tarragon is said (not by me) to be one of the easier herbs to grow. Too far north it is cold sensitive. Annual or monthly fertilizations, e.g. 23:21:17, and regular weedings are recommended.

Leaves can be field stripped as early as June, and, as they are replaced, until frost. Leaves are dried like other herbs. Temperatures over 85 °F may cause discoloration. Foliage yields of about a ton an acre are easily attainable.

USES: Best known for its contribution to tarragon vinegar here in the United States, tarragon is indispensible to French cuisine. Improving a variety of vegetable recipes (e.g. artichoke, asparagus, avocado, bean, beet, cabbage, carrot, celeriac, celery, cucumber, kraut, lettuce, mushrooms, peas, potatoes, rice, salsify, spinach, squash, tomato) and meat dishes involving chicken, crab, duck, fish, lamb, liver, lobster, pheasant, rabbit, salmon, scallops, shrimp, steak, turkey, and veal. Tarragon is also important in soups (chicken, fish, mushroom, tomato, and turtle) and salads (asparagus, bean, chicken, cucumber, lettuce, and tomato). Several sauces are dependent on or indebted to tarragon (bearnaaise, hollandaise, mayonnaise, mousseline). I have successfully used tarragon instead of anise in various liqueur and tea recipes. One delicious liqueur resulted from two spoons of tarragon boiled in sugar water (½ cup), and mixed with ½ pint goat's milk, and 2 oz vodka, served cold. I would not hestitate to substitute tarragon for anise, even for fennel in any recipe.

FOLK MEDICINE: Regarded as anthelmintic, aperient, carminative, diuretic, emmenagogue, febrifuge, hypnotic, refrigerant, sedative, stimulant, stomachic, tonic, and vermifuge, tarragon is said to be useful for amenorrhea, dogbites, stings, stomach, swellings, toothache, and tumor. Tarragon is a promising source of rutin, which reduces capillary permeability and is said to protect against cancer, radiation damage, and strokes. Rutin was once official in the U.S. for allergies, arteriosclerosis, diabetes, and hypertension. Esdragole, on the other hand, may induce cancer in experimental animals.

THYME
(Thymus vulgaris L.)

CULTURE: Thyme, a *perennial*, can be grown on rather sterile rocky soils and may develop its best aromas under such circumstances. Seed (ca 6 lbs / acre) can be hand dibbled or drilled in rows 18 to 36 inches apart, covered by a quarter inch of finely pulverized soil. Volunteers can be planted 10-24 inches apart in new rows or beds. Established plants can be subdivided by root divisions, cuttings, or layering. Densities of 20,000 to 55,000 plants per acre are suggested. Old plants should be dug up and subdivided after about three years, as the woody old plants produce few tender shoots.

Weeding is recommended but fertilization and irrigation recommendations vary. Thyme can take a beating and sometimes seems to profit thereby. Leaves can be stripped off the living plant in place and dried, or the plant may be cut near soil line and tied up in shocks. The leaves may be processed like basil. Dried thyme is said to have a very penetrating aroma. Usually only the flowery tops are harvested, but it is not uncommon to harvest twice in a season. An acre yields ca 1-2 tons fresh thyme. A ton of dry herb is clearly possible.

USES: Perhaps best known here as a seasoning for chicken, thyme is also used to flavor or scent butters, cheeses, chowders, fish, liqueurs, meats, olives, onions, perfumes, pickles, poultry, sauces, soaps, soups, stews, stuffings, tomatoes, and vinegars. Thyme is used to season many vegetables (beans, both dry and fresh; beets, carrots, celery, cucumber, eggplant, leeks, mushrooms, onions, peas, peppers, potatoes, pumpkins, spinach, squash) and meats (beef, boar, brain, crab, dove, duck, fish, goose, grouse, lamb, lobster, mussel, oxtail, pheasant, pork, rabbit, sausage, scallop, shrimp, veal, and venison). Thyme can serve as salt or pepper for chicken. It is an ingredient of the liqueur Benedictine. Thyme is one of the better honey plants. Sheep fed on thyme are said to develop an especially delicate flavor. Currently, thyme is popular as a tisane, mixed with such herbs as mint and rosemary. Thyme tea (1 teaspoon fresh leaves steeped in one cup boiling water) is said to be good for headache. Thyme butter spread over steak before cooking is good.

THYME

FOLK MEDICINE: Medicinally, thyme has been used in anemia, asthma, bronchitis, bruises, catarrh, cold, colic, coughs, cramps, diabetes, diarrhea, dysmenorrhea, fever, flatulence, flu, gout, gingivitis, halitosis, hangover, headache, heart, hysteria, indigestion, infection, insomnia, itch, laryngitis, leprosy, leucorrhea, melancholy, nerves, neuralgia, nightmares, rheumatism, scarlet fever, sciatica, snakebite, sores, sore throat, spleen disorders, stomach, stomatitis, uterine disorders, warts, and whooping cough. It is considered antiseptic, antispasmodic, aperient, astringent, carminative, diaphoretic, digestive, ecbolic, emmenagogue, fungicidal, nervine, rubefacient, tonic, and vermifuge. Smoked like tobacco, it is used for digestion, drowsiness, and headache. Burning thyme is supposed to repel insects. Dried flowers, like lavender are said to preserve linens from insects, scenting them at the same time. USDA research has not confirmed this suggestion.

Thymol, the oil of thyme, is an antiseptic and deodorant, used externally and internally. It has been used both as a meat preservative, and to treat burns, eczema, psoriasis, worms, and ringworm. It is used as a fungicide to prevent mildew, and as toothpaste ingredient. It is used as a diffusible stimulant in case of collapse. It is said to cause mental excitement. It has been used externally as a rubefacient and counterirritant. Thymol is sometimes suggested for blastomycoses, coccidioses, and moniliasis. Carvacol and thymol have tracheal properties. The antispasmodic action of thymol is probably due to the phenols. I take tisanes of thyme-scented herbs for lower back episodes.

While most of the herbs herein discussed are classified as GRAS (Generally Recognized As Safe) by the FDA, essential oils derived and concentrated from these herbs can be dangerous. Thymol, for example, can cause fatalities.

CHAPTER 2

Herbal Zest

That America is interested in herbal zest is indicated in Table 3. US imports of specified condiments, seasons and flavoring materials (see Table 3) in 1979 were nearly 150,000 tons worth $200 million, only slightly lower than the record 1978 level. U.S. essential oil imports are shown in Table 4. U.S. spice exports in 1979 fell to 10,719 tons worth about $20 million compared with 1978 shipments of 11,357 tons worth $20.3 million. Prices for some of our spice imports are shown in Table 5, more recent import figures in Table 6.

Just before Time Inc. closed it, the Washington Star issued its 81 Cookbook Part 2 (July 26, 1981) consisting of winning recipes from the readers of the Washington Star. Among the first 100, almost all contained one or more herbal condiments or spices, over and above salt and pepper (black pepper in general considered a spice not an herb). More than 40 different

Table 3. US Condiment Imports (1979)[1]

Common Name	Scientific Name	Import (Kilograms)	Total Value ($)	Value per Kilogram ($)
allspice	Pimenta dioica (L.) Merr.	487,300	817,000	1.67
anise	Pimpinella anisum L.	492,200	803,900	1.63
basil	Ocimum basilicum L.	484,500	473,400	0.97
capers	Capparis spinosa L.	657,500	2,606,700	3.96
capsicum	Capsicum spp.	5,190,300	6,273,400	1.20
caraway seed	Carum carvi L.	3,586,000	5,224,700	1.45
cardamom seed	Elettaria cardamomum (L.) Maton	90,600	922,400	10.18
cassia	Cinnamomum aromaticum Nees	9,079,800	6,706,000	0.73
celery seed	Apium graveolens L.	2,149,500	1,577,300	0.73
cinnamon	Cinnamomum verum J.S. Presl	472,800	867,700	1.83
cloves	Syzygium aromaticum (L.) Merr. & Perry	1,321,000	9,091,700	6.88
coriander	Coriandrum sativum L.	3,300,700	1,273,200	0.38
cumin seed	Cuminum cyminum L.	5,802,900	10,626,200	1.83
curry		161,800	455,800	2.81
dill seed	Anethum graveolens L.	556,500	368,800	0.66
fennel seed	Foeniculum vulgare Mill.	1,157,800	933,500	0.80
dehydrated garlic	Allium sativum L.	196,200	362,000	1.84
ginger	Zingiber officinale Roscoe	263,300	459,000	1.74

leaves of laurel	Laurus nobilis L.	371,500	574,900	1.54
mace	Myristica fragrans Houtt.	250,300	523,100	2.08
marjoram	Origanum majorana L.	322,700	317,700	0.98
mint leaf	Mentha spicata L.	250,100	626,000	2.50
mustard seed	Brassica spp.	28,676,100	10,727,400	0.37
nutmeg	Myristica fragrans Houtt.	2,406,200	4,697,800	1.95
dehydrated onions	Allium cepa L.	25,100	40,300	1.60
origanum leaves	Origanum spp.	2,693,900	5,123,700	1.90
paprika	Capsicum sp.	5,567,700	8,325,300	1.49
parsley	Petroselinum crispum (Mill.) Nym. ex A.W. Hill	147,300	244,600	1.66
pepper (black)	Piper nigrum L.	24,482,400	42,546,500	1.73
pepper (white)	Piper nigrum L.	2,755,600	6,828,900	2.47
poppyseed	Papaver somniferum L.	2,364,400	1,422,700	0.61
rosemary	Rosmarinus officinalis L.	245,600	157,500	0.64
sage	Salvia officinalis L.	1,471,000	3,688,800	2.50
savory	Satureja spp.	134,000	258,200	1.92
sesame seed	Sesamum indicum L.	32,099,300	31,648,000	0.98
tarragon	Artemisia dracunculus L.	40,600	215,900	5.31
thyme	Thymus vulgaris L.	815,200	1,358,600	1.66
turmeric	Curcuma domestica Val.	1,539,800	1,684,300	1.09
vanilla beans	Vanilla planifolia Andr.	496,700	18,291,000	36.82
mixed spices		975,000	1,427,400	1.46

[1] Based on Duke (1982).

Table 4. US Essential Oil Imports (1979)[1]

Common Name	Scientific Name	Import (Kilograms)	Total Value ($)	Value per Kilogram ($)
bitter almond oil	Prunus dulcis (Mill.) D.A. Webb	54,133	187,000	3.59
anise oil	Pimpinella anisum L.	34,049	543,800	15.97
bergamot oil	Citrus sp.	29,424	1,558,800	52.98
oil of camphor	Cinnamomum camphora (L.) J.S. Presl	37,692	91,600	2.43
oil of caraway	Carum carvi L.	10,349	327,800	31.68
oil of cassia	Cinnamomum aromaticum Nees	68,397	3,907,500	57.13
oil of cedar leaf	Juniperus spp.	7,724	276,000	35.73
oil of cedarwood	Juniperus spp.	47,180	93,900	1.99
oil of cinnamon	Cinnamomum verum J.S. Presl	26,474	165,400	6.25
oil of citronella	Cymbopogon nardus (L.) Rendle	645,642	3,474,200	5.38
oil of citrus	Citrus spp.	22,079	765,500	34.67
oil of clove	Syzygium aromaticum (L.) Merr. & Perry	781,761	4,300,600	5.50
oil of cornmint	Mentha arvensis L.	208,349	1,792,700	8.60
oil of eucalyptus	Eucalyptus spp.	277,456	916,100	3.30
oil of geranium	Pelargonium spp.	62,570	3,869,200	61.84
oil of grapefruit	Citrus paradisi Macfad.	27,387	58,500	2.14
oil of lavender	Lavandula spp.	140,125	3,325,500	23.73
oil of lemon	Citrus limon (L.) Burm. f.	673,264	10,755,600	15.98

oil of lemongrass	Cymbopogon spp.	179,698	1,214,400	6.76
oil of lignaloe or bois de rose	Ocotea sp?	83,985	921,000	10.97
oil of lime	Citrus aurantiifolia (Christm.) Swingle	813,836	20,966,000	25.76
oil of neroli or orange flower	Citrus aurantium L.	162	159,500	984.62
oil of nutmeg	Myristica fragrans Houtt.	151,591	2,740,400	18.08
oil of onion and garlic	Allium cepa L. and Allium sativum L.	4,213	594,900	14.12
oil of orange	Citrus sp.	2,755,290	2,229,000	0.81
oil of origanum	Origanum sp.	6,385	198,600	31.10
oil of orris	Iris x germanica L.	351	351,800	1,002.16
oil of palmarosa	Cymbopogon martinii (Roxb.) W. Wats.	12,758	298,500	23.39
oil of patchouli	Pogostemon cablin (Blanco) Benth.	237,949	5,857,000	24.61
oil of peppermint	Mentha x piperita L.	3,418	99,800	29.20
oil of petitgrain	Citrus sp.	159,145	1,834,600	11.53
oil of pine	Pinus sp.	1,992	29,200	14.64
oil of pineneedle	Pinus sp.	12,986	207,600	15.98
oil of attar of rose	Rosa sp.	981	1,857,800	1,893.74
oil of rosemary	Rosmarinus officinalis L.	60,359	742,000	12.29
oil of sandalwood	Santalum album L.	32,716	3,169,700	96.89
oil of sassafras	Sassafras albidum (Nutt.) Nees	283,642	948,400	3.34
oil of spearmint	Mentha spicata L.	1,885	28,100	14.92
oil of thyme	Thymus vulgaris L.	8,589	241,300	28.09
oil of vetiver	Vetiveria zizanioides (L.) Nash ex Small	78,744	4,439,700	56.38
oil of ylang ylang	Cananga odorata (Lam.) Hook.fil. & Thoms.	50,410	2,358,400	46.78
other essential oils		936,210	18,406,100	19.66

[1] Based on Duke (1982).

Table 5.
Specified Condiments and Flavoring Materials: Approximate New York Spot Prices As Of Early March, 1981-1983
(In cents per pound)

Item	1981	1982	1983
ALLSPICE (PIMENTO):			
Guatemalan............................:	100	93	114
Honduran...............................:	100	93	114
Jamaican ....:	107	119	117
Mexican.................................:	80	78	(¹)
ANISE:			
Chinese star............................:	185	170	155
Egyptian ....:	110	135	79
Spanish.................................:	139	239	117
Turkish ....:	118	140	79
BASIL:			
Egyptian ....:	50	97	90
French..................................:	(¹)	97	105
Domestic ....:	285	310	375
CAPSICUM PEPPERS:			
Chinese ....:	76-90	88-93	75
Indian ....:	74	77	70
Pakistan................................:	59	85	66
CARAWAY:			
Dutch ....:	59	71	56
Egyptian ....:	50	49	50
CARDAMOM:			
Bleached "A"...........................:	1200	1200	1375
Decorticated...........................:	410	335	355
Guatemalan fancy greens ....:	(¹)	950	775
Guatemalan mixed greens ....:	400	350	315
CASSIA:			
Chinese, Taiwan......................:	40	(¹)	43
Indonesian Batavia "AA" ....:	132	130	135
Indonesian Korintje "AA" ....:	90	68	62
CELERY SEED:			
Indian ....:	42	43	43
CINNAMON:			
Ceylon No. 2 ....:	135	130	135
Seychelles.............................:	59-62	42	50
CLOVES:			
Brazilian................................:	420	515	445
Madagascar...........................:	445	530	470

Table 5. (cont.)

CORIANDER:			
Argentine.............................:	30	34	(¹)
Chinese ....:	26	34	(¹)
Egyptian ....:	24	(¹)	(¹)
Moroccan.............................:	30	37	37
Romanian ....:	28	36	35
CUMIN:			
Indian..................................:	76	85	92
Turkish ....:	(¹)	(¹)	85
Pakistan...............................:	76	78	87
DILL SEED:			
Dewhiskered ....:	50	44	36
FENNEL:			
Egyptian ....:	42	42	68
Indian..................................:	65	63-72	86
FENUGREEK:			
Indian..................................:	39	38	34
Moroccan.............................:	39	(¹)	(¹)
GINGER:			
Chinese, whole peeled ....:	50	48	57
Chinese, sliced........................:	46	43	48
Indian, cochin ....:	54	69	110
Jamaican No. 3 ....:	190	215	138
LAUREL (BAY) LEAVE:			
Turkish ....:	94	55	64
Turkish cutting grade ....:	89	52	54
MACE:			
East Indian ....:	145	175	225
MARJORAM:			
Egyptian ....:	60	65	150
French...................................:	82-95	92	165
MINT LEAVES:			
Peppermint ....:	250	250	240
Spearmint ....:	225	235	225
MUSTARD SEED:			
Canadian No. 1 yellow..............:	26	24	23
Oriental ....:	23	23	21
NUTMEGS:			
East Indian ....:	91	89	80
West Indian ....:	130	95	80
ORIGANUM:			
Greek 30 mesh ....:	123	145	125
Mexican................................:	70	115	108
Turkish 30 mesh ....:	102	130	115

Table 5. (cont.)

PAPRIKA:
Spanish 90 ASTA..................:	106	99	95
Spanish 120 ASTA.................:	(¹)	115	98
Spanish 100 ASTA.................:	110	102	92

PARSLEY:
Domestic ....:	160	160	160
Imported...............................:	125	145	75

PEPPER, BLACK:
Brazilian................................:	66	69	54
Indonesian, Lampong ....:	76	73	64
Indian, Malabar ....:	76	75	64
Sarawak................................:	82	76	66

PEPPER, WHITE:
Brazilian................................:	95	93	85
Indonesian, Muntok.................:	97	93	85

POPPY SEED:
Australian ....:	48	80	85
Dutch....................................:	51	82	78-88
Turkish..................................:	44	76	72

ROSEMARY:
French...................................:	68	57	(¹)
Portuguese............................:	46	46	46
Spanish..................................:	46	46	46

SAFFRON:
Spanish..................................:	46000	41000	29000

SAGE:
Albanian ....:	185	140	112
Dalmation No. 1......................:	215	185	160
Turkish..................................:	68	85	72

SAVORY:
Albanian ....:	90	50	(¹)
French...................................:	110	60	(¹)
Yugoslavian...........................:	110	60	55

SESAME SEED:
Guatemalan natural ....:	44	52-62	68
Nicaraguan hulled ....:	63	63	(¹)

TARRAGON:
Domestic ....:	900	900	900

THYME:
French...................................:	153	129	118
Spanish..................................:	105	94	78

TURMERIC:
Haitian...................................:	40	29	34
Indian, Alleppey......................:	43	39-43	52

Table 5. (cont.)

VANILLA BEANS:

Madagascar, bourbon ....:	3150	2800	3000-3100
Indonesia, Java.......................:	2850	2500	2700-3000

[1] Quotations not available.

SOURCE: FAS/USDA. Apr, 1983

51

Table 6.
United States: Imports of Specified Condiments, Seasonings, and Flavoring Materials, 1981 and 1982

Condiments and flavoring materials[1]	1981		1982	
	Metric tons	1,000 dollars	Metric tons	1,000 dollars
Allspice (pimento)	813.5	1,436.9	523.4	1,123.2
(ground)	2.0	3.3	.5	1.2
Anise seed	524.2	1,266.0	619.5	1,631.2
Basil	744.9	684.2	1,119.2	1,353.7
(other than crude)	3.2	17.0	10.4	27.8
Capers	757.5	3,709.7	166.7	777.4
Capsicum or red peppers:				
Anaheim and ancho	796.0	1,530.4	1,220.1	1,772.2
Other	4,350.9	5,321.5	4,525.2	6,049.3
(ground)	171.7	383.4	156.2	358.1
Caraway seed	3,031.2	3,162.3	3,590.5	4,035.8
Cardamom seed	83.7	491.0	119.0	785.2
Cassia	8,173.1	9,466.8	8,284.7	9,024.5
(ground)	269.4	245.5	428.2	416.6
Celery seed	2,040.9	1,371.0	1,959.0	1,326.0
Cinnamon	844.4	1,418.2	816.6	1,454.4
(ground)	44.2	115.6	54.2	125.2
Cloves	944.2	8,096.3	1,105.8	10,997.9
(ground)	.3	1.9	.9	7.6

52

Coriander	4,663.4	1,852.8	4,491.5	1,867.4
Cumin seed	4,726.6	6,115.0	4,031.9	5,993.7
Curry and curry powder	221.0	749.5	277.8	976.9
Dill seed	537.7	498.8	603.0	517.4
Fennel seed	1,416.0	1,268.3	1,379.9	1,287.2
Garlic (dehydrated)	118.9	155.7	1,597.0	2,098.5
Ginger	4,324.5	3,859.8	4,752.0	4,164.3
(ginger)	54.1	66.0	53.5	87.1
(sweet)	54.7	110.2	108.3	184.7
(candied)	181.8	460.6	143.0	352.2
Laurel (Bay) leaves	394.0	654.1	467.1	592.5
(other than crude)	1.5	3.0	3.0	7.4
Mace	261.8	532.4	215.4	595.1
(ground)	5.2	10.0	8.4	23.5
Marjoram	343.0	371.6	514.2	622.0
(other than crude)	10.9	11.7	.4	.9
Mint leaves	214.7	301.1	226.4	290.1
(manufactured)	34.3	175.5	82.1	268.2
Mustard seed	35,091.2	10,984.8	31,655.0	11,549.0
(ground)	1,123.8	1,806.9	1,228.9	1,753.5
(other)	1,118.2	2,169.8	1,310.1	2,423.0
Nutmegs	2,170.3	3,525.6	2,421.2	3,576.9
(ground)	32.2	80.3	25.4	50.2
Onions (dehydrated)	99.6	101.7	11.9	34.4
Origanum leaves	3,067.9	5,118.7	3,792.3	6,917.6
(other than crude)	56.3	68.9	300.2	247.4
Paprika³	4,499.5	7,668.0	4,089.1	6,748.7
Parsley	137.9	145.4	284.2	284.6
(manufactured)	44.6	132.5	73.4	197.1
Pepper, black	28,563.4	36,327.9	27,811.0	30,504.5

Table 6. (cont.)

Pepper, white	2,503.1	4,455.0	2,720.9	4,282.1
(black and white ground)	50.7	153.3	81.6	218.0
Poppy seed	2,842.1	2,336.2	3,313.7	4,231.2
Rosemary	267.0	185.0	275.0	183.8
(other than crude)	1.8	8.8	.4	4.3
Sage	1,481.7	4,827.6	1,444.4	3,613.0
(ground or rubbed)	14.8	56.3	11.7	34.7
Savory	54.1	60.5	59.1	43.4
(other than crude)	.2	.9	(2)	1.3
Sesame seed	37,954.0	35,202.3	33,213.0	32,454.4
Tarragon	29.6	286.5	60.1	408.4
(other than crude)	6.9	35.1	6.5	30.6
Thyme	577.9	1,043.6	702.3	1,137.5
(other than crude)	10.3	25.4	2.4	8.5
Turmeric	1,862.5	1,129.6	1,604.2	946.8
Vanilla beans	640.2	31,248.2	883.8	45,196.1
Mixed spices	847.0	1,860.2	1,255.6	2,397.7
Total	166,308.2	206,962.1	162,292.4	220,675.1

[1] Unground, unless otherwise specified.   [2] 50 kilograms or less.   [3] Ground and unground.

NOTE: All values refer to f.o.b. country of origin.

Source: FAS/USDA. Apr, 1983

herbs and botanical condiments were cited in the first hundred recipes, and several have five or more herbs in a single recipe. On the other hand curry and fivespice represent combinations of several herbal products.

Allspice occurred in only one recipe of the first hundred, (meatpie); basil in 5 (artichoke, chicken, spaghetti, veal, zucchini); bay in 4 (bean, lamb, shrimp, zucchini); capers in 2 (artichoke, crab); caraway in 2 (liptauer, vegetable dish); cayenne in 4 (mushroom, cheeseball, gumbo, salmon); celery in more than 10 (beef, chicken, crab, gumbo, lamb, meatball, sausage, shrimp, zucchini); celery salt in 2 (celery, crab); chives in 4 (giblets, lobster, shrimp, vegetable), cinnamon in 4 (chicken, cranberry, hamburger, veal); cloves in 3 (chicken, cranberry, ham); cumin in 2 (chicken, tofu); curry in 3 (crab, curry, spinach), fennel in 1 (veal); file in 3 (gumbo, rice); fivespice in 1 (chicken); garlic in more than 35 (artichoke, bean, chicken, crab, fish, guacamole, gumbo, ham, lamb, liver, meatballs, mushrooms, pate, porkchop, seafood, seafood creole, shrimp, steak, tahini, vegetable veal, vegetarian lasagna, and zucchini); ginger in 4 (chicken, porkchop), hickory chips in 1 (porkchop); horseradish in 4 (chicken, salmon, shrimp); juniper berries in 1 (porkchop); marjoram in 2 (liver, porkchop); mint in 1 (hamburger); mustard in more than 10 (chicken, crab, ham, hamloaf, liptauer, liver, omelet, porkchop, quiche, shrimp, sloppyjoe); nutmeg in 3 (giblets, lamb, veal); onion in more than 35 (artichoke, bean, beefpie, beefroast, brisket, cabbage, caviar, celery, cheese, cheeseball, chicken, chile, crab, curry, eggplant, guacamole, gumbo, hamburger, lamb, liptauer, liver, meatball, meatpie, mushroom, omelet, pork, porkchop, salmon, sausage, seafood, shrimp, sloppyjoe, spinach, stew, tofu, veal, vegetable, vegetarian dishes, and zucchini); paprika in more than 10 (bean, celery, chicken, liptauer, omelet, pork, seafood, shrimp, sparerib); parsley in 25 (artichoke, bean, beefpie, chicken, cornish hen, crab, curry, fish, gumbo, liver, lobster, meatball, mushroom, lamb, pork, porkchop, salmon, scallop, shrimp, spaghetti, vegetarian lasagna, zucchini); pepper (green) in 7 (cheeseball, crab, gumbo, omelet, seafood creole, shrimp, veal); pepper (chili, hot, or jalapeno); in 10 (artichoke, brisket, chicken, liver, shrimp, sparerib, tofu, vegetable); pepper (red) in 3 (gumbo, liver, shrimp), pimento in 2 (artichoke, cheeseball); rosemary in 3 (lamb, scallops, seafood creole); sage in 1 (porkchop); scallions in 5 (artichoke, mushroom, quiche, seafood, spaghetti); sesame in 1 (chicken); shallot in 1 (mushroom); tabasco in 10 (artichoke, brisket, cheese, chicken, crab, guacamole, gumbo, ham, pate, shrimp, zucchini); tarragon in 4 (chicken, shrimp, vegetable dip); and thyme in 9 (artichoke, crab, gumbo, liver, pate, porkchop, scallop, vegetarian lasagna, zucchini). With an herb in the majority of recipes, the value of herbs to American consumers and small farmers is indicated.

So you finally have a good garden full of herbs. What are you going to do with them? Here's how some of them are being used. I group the herbs according to the fruit and vegetable, meat and speciality dishes in which they are frequently reported to be used.

Fruits and Vegetables

APPLE PIE: cardamon, cinnamon.
APPLES: anise, basil, caraway, cardamon, celery, mint, pineapplement.
ARTICHOKE: basil, capers, chile, dill, garlic, onion, oregano, parsley, pimento, scallion, tabasco, thyme.
ASPARAGUS: balm, burnet, chervil, fennel, oregano, parsley, savory, sorrel, tarragon.
ASPARAGUS SALAD: chervil, parsley, tarragon.
AVOCADO: marjoram, oregano.
BAKED BEANS: rosemary, savory.
BEAN (DRY): basil, bay, burnet, curry, garlic, juniper, mint, onion, oregano, paprika, parsley, rosemary, savory, thyme.
BEAN (FRESH GREEN): basil, caraway, clove, coriander, dill, fennel, oregano, savory, sorrel, thyme.
BEAN SALAD: chive, parsley, sorrel.
BEAN SOUP: burnet, savory.
BEET: anise, caraway, clove, dill, horseradish, tarragon, thyme.
BEET AND CUCUMBER SALAD: chive, parsley, tarragon.
BEET SALAD: caraway, chives, coriander, fennel, lovage, mint, parsley, thyme.

BROADBEAN: savory.

BROADBEAN SALAD: lemonthyme, parsley, savory.

BROCCOLI: basil, horseradish, marjoram, oregano.

BRUSSELS SPROUTS: marjoram.

CABBAGE: allspice, anise, caraway, cardamon, cumin, dill, fennel, juniper, marjoram, onion, poppyseed, savory, sorrel, tarragon, thyme.

CABBAGE SALAD: caraway, chives, dill, lovage, marjoram, nasturtium, parsley, poppyseed, sorrel, thyme, watercress.

CARAWAY ROOT: nettle, parsley, sorrel, tarragon (Norway).

CARROT: allspice, anise, basil, caraway, celery, chive, cicely, cinnamon, clove, garlic, lovage, marjoram, mint, oregano, parsley, shallot, tarragon, thyme.

CARROT SALAD: anise, bay, chives, cicely, fennel, marjoram, mint, parsley, thyme.

CAULIFLOWER: caraway, fennel, marjoram, rosemary, savory, sorrel.

CAULIFLOWER SALAD: basil, marjoram, mint, parsley.

CELERIAC AND POTATO SALAD: bay, parsley, sorrel, thyme, watercress.

CELERY: bay, burnet, chives, onion, paprika.

CELERIAC: basil, parsley, thyme.

CHICORY: marjoram, tarragon.

CORN: chervil, chile, chives, coriander, parsley.

COS LETTUCE: basil, marjoram.

CRANBERRY: cinnamon, clove.

CUCUMBER: anise, basil, bay, caraway, coriander, dill.

CUCUMBER SALAD: chive, dill, fennel, nasturtium, watercress.

DANDELION SALAD: borage, chives, garlic, parsley.

EGGPLANT: allspice, basil, chile, cinnamon, fennel, garlic, marjoram, mint, onion, oregano, sage.

ENDIVE SALAD: chives, garlic, parsley, sorrel, watercress.

FENNEL: chives.

FRUIT SALAD: anise, balm, caraway, cardamon, cicely, celeryseed, cinnamon, clove, coriander, ginger, mint.

GRAPEFRUIT: rosemary.

JERUSALEM ARTICHOKE: balm, thyme.

KOHLRABI: lovage, thyme.

KRAUT: caraway, cumin, dill, fennel, juniper, mustard.

LEEKS: parsley, thyme.

LENTILS: fennel seed, savory, sorrel.
LENTIL SOUP: chervil, lovage, savory, sorrel.
LETTUCE: chives, sorrel.
LETTUCE AND TOMATO SALAD: basil, chervil, parsley, savory, tarragon.
LETTUCE SALAD: chervil, chives, lavender, mint, parsley, tarragon.
LIMA BEAN: sage
MELON: cardamon.
MUSHROOM: balm, basil, cayenne, chive, fennel, garlic, lovage, marjoram, mint, onion, oregano, parsley, rosemary, scallion, shallot, tarragon, thyme.
MUSHROOM SOUP: burnet, fennel, parsley, tarragon.
NASTURTIUM-CUCUMBER SALAD: dill, fennel, poppy-seed.
ONION: parsley, rosemary, sage, thyme.
ONION SOUP: marjoram, sage, thyme.
PEACH: cardamon.
PEAR: caraway, marjoram, oregano.
PEAS: basil, chives, fennel, marjoram, parsley, mint, poppy-seed, rosemary, savory, tarragon.
PEA SOUP: costmary, mint, parsley, savory, tarragon, thyme.
PEPPERS: chives, oregano.
PINEAPPLE: mint, pineapple mint.
POTATOES: basil, caraway, chervil, chives, dill, fennel, lovage, marjoram, mint, parsley, rosemary, sage, savory, sorrel, tarragon, thyme.
POTATO SALAD: caraway, celery seed, chives, cicely, dill, garlic, mint, parsley, savory, sorrel, watercress.
POTATO SOUP: dill, marjoram, mint, nasturtium, parsley, watercress.
PUMPKIN: cardamon, cinnamon.
RADISH: chives.
RHUBARB: angelica, cicely.
RICE: marigold, saffrom, sassafras, tumeric.
RUTABAGA: lovage, marjoram, poppyseed.
SPINACH: basil, curry, marjoram, mint, peppermint, rosemary, sorrel.
SQUASH: allspice, basil, bay, cardamon, cinnamon, dill, marjoram, mint, parsley, sage, thyme.
SWEETPOTATO: allspice, cardamon, clove.

TOMATO: allspice, basil, bay, celeryseed, cinnamon, clove, dill, marjoram, oregano, sage, sorrel, thyme.
TOMATO AND CHEESE SALAD: chives, nasturtium, parsley, watercress.
TOMATO JUICE: basil, celery, lovage, tarragon
TOMATO SALAD: basil, chives, oregano, parsley, savory, tarragon.
TOMATO SAUCE: basil, bay, celery, clove, marjoram, oregano, parsley.
TOMATO SOUP: basil, chives, lavender, marjoram, nasturtium, oregano, parsley, rosemary, tarragon, thyme, watercress.
TURNIPS: caraway, poppyseed, rosemary, savory.
VEGETABLE SALAD (RAW): caraway, chile, lavender, lovage, parsley, poppyseed, savory, sorrel, watercress.
VEGETABLE SOUP: balm, bay, caraway, chervil, chives, cicely, coriander, lavender, marjoram, parsley, sorrel, thyme, watercress.
WATERCRESS APPLE SALAD: dill, fennel, marjoram.
ZUCCHINI: bay, basil, caraway, celery, dill, garlic, mint, onion, oregano, parsley, rosemary, savory, tabasco, thyme.

Meats

BACON: chive, sage, shallot.
BEEF: allspice, anise, bay, celery, chervil, chives, clove, cumin, dill, garlic, horseradish, lemonthyme, marjoram, mint, onion, paprika, parsley, rosemary, sage, savory, thyme.
BEEF STEW: balm, basil, chervil, cicely, coriander, hyssop, lavender, marjoram, onion, rosemary.
BOILED CHICKEN: coriander, lemonthyme, marjoram, savory, sorrel, tarragon.
BRISKET: allspice, chile, onion, parsley, tabasco.
CACCIATORE: basil, garlic, oregano, parsley, pepper.
CAVIAR: onion.
CHICKEN: balm, basil, cayenne, celery, chervil, cinnamon, coriander, cumin, costmary, dill, fennel, garlic, ginger, horseradish, lemonthyme, lovage, marjoram, mustard, nutmeg, onion, oregano, paprika, parsley, rosemary, saffron, sage, savory, sesame, tabasco, tarragon, thyme.
CHICKEN SOUP: balm, chervil, lavender, lemonthyme, lovage, parsley, rosemary, sorrel, tarragon.

CLAM: thyme.
CORNED BEEF: cinnamon, horseradish, mustardseed.
CORNISH HEN: parsley.
CRAB: capers, celery, curry, fennelseed, garlic, mustard, onion, parsley, sage, tabasco, tarragon, thyme.
CRAYFISH: dill.
DEER: coriander.
DOVE: parsley, sage, shallot, thyme.
DUCK: chives, clary, coriander, marjoram, pepper, rosemary, saffron, sage, sorrel, staranise, thyme.
EEL: basil, clary, fennel, sage, sorrel.
FISH: allspice, angelica, anise, balm, basil, bay, cardamon, cayenne, celery, chervil, chive, cicely, clary, clove, cumin, dill, fennel, garlic, ginger, horseradish, hyssop, lemonthyme, marjoram, mint, oregano, paprika, parsley, rosemary, saffron, sage, savory, sesame, sorrel, tarragon, thyme, tumeric, watercress.
FISH SALAD: chervil, chives, dill, fennel, parsley, watercress.
FISH SOUP: bay, chervil, dill, fennel, garlic, parsley, savory, thyme.
GOOSE: caraway, clary, costmary, marjoram, mugwort, saffron, sage, shallot, sorrel, thyme.
GROUSE: parsley, sage, shallot, thyme.
HAM: allspice, basil, chive, cinnamon, clove, dill, horseradish, lemonthyme, marjoram, mustardseed, sage, shallot, tabasco.
HAMBURGER: cinnamon, mint, onion.
HERRING: bay, capers, nasturtium, parsley, thyme.
LAMB: balm, basil, bay, caraway, celery, cinnamon, coriander, dill, fennelseed, garlic, lavender, marjoram, mint, nutmeg, onion, oregano, parsley, rosemary, sage, southernwood, tarragon, thyme.
LAMBCHOP: basil, dill, marjoram.
LAMB STEW: bay, chervil, coriander, costmary, lavender, marjoram, parsley, rosemary, thyme.
LIVER: allspice, basil, bay, chile, garlic, marjoram, mustard, onion, parsley, sage, thyme.
LOBSTER: chives, parsley, tarragon.
MACKEREL: basil, clary, dill, fennel, marjoram, sage, tarragon.
MEATBALL: cardamon, celery, garlic, onion, parsley.
MEATLOAF: celeryseed, coriander, parsley.

MEATPIE: allspice, onion.

MEAT SOUP: balm, bay, celeryseed, chervil, chives, coriander, lavender, marjoram, parsley, rosemary, sorrel.

MUSSELS: chervil, sorrel, tarragon, thyme.

MUTTON: bay, dill, fennelseed, garlic, lavender, parsley, rosemary.

OXTAIL: caraway.

PARTRIDGE: basil, rosemary.

PATE: bay, garlic, onion, tabasco, thyme.

PHEASANT: basil, bay, marjoram, oregano, rosemary.

PORK: balm, bergamot, caraway, chive, cinnamon, clary, clove, coriander, cumin, dill, marjoram, mugwort, oregano, paprika, parsley, rosemary, saffron, sage, shallot, sorrel, staranise, thyme.

PORK CHOP: dill, garlic, ginger, juniper, marjoram, mustard, onion, parsley, sage, thyme.

POULTRY: anise, balm, basil, cardamon, celery, clary, coriander, dill, garlic, hyssop, lovage, marjoram, oregano, paprika, parsley, rosemary, saffron, sage, savory, sesame, tarragon, thyme, tumeric.

RABBIT: basil, bay, lemonthyme, sage, thyme.

RISSOLES: basil, chive, marjoram, parsley.

ROAST BEEF: bay, caraway, garlic, lavender, marjoram, onion, rosemary, thyme.

ROAST CHICKEN: lavender, rosemary, tarragon, thyme.

ROAST PORK: basil, caraway, coriander, dill, fennelseed, garlic, lavender, lemonthyme, marjoram, oregano, rosemary, sage, savory, thyme.

ROAST VEAL: basil, costmary, marjoram, oregano, rosemary, tarragon, thyme.

SALMON: cayenne, dill, fennel, horseradish, marjoram, onion, parsley, sorrel, tarragon.

SAUSAGE: basil, caraway, celery, cinnamon, coriander, cumin, fennel, ginger, marjoram, onion, paprika, sage, savory, thyme.

SCALLOPS: chervil, oregano, parsley, rosemary, sorrel, tarragon, thyme.

SEAFOOD: anise, dill, fennel, garlic, horseradish, marigold, nutmeg, onion, paprika, rosemary, saffron, scallion, tarragon, thyme.

SHELLFISH: anise, basil, chervil, dill, horseradish, marigold, onion, parsley, saffron, tarragon.

SHRIMP: basil, bay, celery, chile, chives, dill, fennelseed, garlic, horseradish, mustard, onion, paprika, parsley, tabasco, tarragon.
SNAILS: fennel.
SPARERIBS: caraway, chile, paprika.
STEAK (BEEF): garlic, marjoram, parsley, pepper.
TONGUE: bay, cinnamon, horseradish.
TRIPE: lemonthyme, sage.
TROUT: savory.
TURKEY: basil, lemonthyme, lovage, marjoram, parsley, rosemary, sage, savory, tarragon, thyme.
TURTLE: basil.
VEAL: basil, chervil, cinnamon, costmary, fennel, garlic, marjoram, mint, nutmeg, onion, paprika, sage, tarragon, thyme.
VEAL CUTLET: chervil, costmary, parsley.
VENISON: bay.
WHITE FISH: basil, bay, chive, dill, fennel, marjoram, parsley, tarragon.

Miscellaneous

ASPIC: bay (I've used dozens of herbs with pleasure with aspics).
BAKED EGG: basil, chervil, chive, costmary, parsley, tarragon, thyme.
BARBEQUE SAUCE: basil, chile, cumin, garlic, oregano.
BEER INGREDIENTS: barley, birch, ginger, hops, oats, rice, scotchbroom, sweetflag, wheat.
BISCUITS: chive, herb-cheese, marjoram, mint, rosemary, sage, tansy, thyme.
BOQUET GARNI: bay, marjoram, parsley, savory, thyme.
BREAD: anise, caraway, cinnamon, clove, cumin, dill, fennel, fenugreek, ginger, marjoram, oregano, poppyseed, rosemary, saffron, sesame, thyme.
BUTTER: chervil, chive, dill, garlic, horseradish, marigold, mint, nasturtium, oregano, parsley, rue, sage, savory, sesame, shallot, tarragon, thyme, watercress.
CAKE: allspice, basil, caraway, cardamon, coriander, fennel, ginger, lovage, mace, nutmeg, poppyseed, sesame.
CANAPES: chervil, chive, fennel, herb-cheese, tarragon.

CANDY: lovage, peppermint, perilla, sesame.

CHEESE: basil, burnet, caraway, cayenne, chervil, chile, chive, cumin, dill, fennel, garlic, lavender, lovage, marigold, marjoram, mint, nasturtium, onion, oregano, parsley, poppyseed, rosemary, rue, sage, tansy, thyme, watercress.

CHILE SAUCE: coriander, cumin.

CHOWDERS: dill, saffron, sage, thyme.

CHUTNEY: clove, coriander, mustardseed.

COFFEE: cardamon.

CONSERVES: lavender, mint, rose, rosemary, violet.

COOKIES: allspice, anise, cardamon, clove, fennel, ginger, mace, nutmeg, poppyseed, sesame.

CURRIES: cardamon, cayenne, celery, clove, coriander, cumin, dill, fenugreek, ginger, mace, onion, parsley, pepper, tumeric.

EGGS: basil, celery, chervil, chive, dill, garlic, parsley, savory.

EGG SALAD: chives, parsley, sorrel, tarragon, thyme, watercress.

FINES HERBS: basil ± chervil ± chives ± fennel ± marjoram ± parsley ± rosemary ± sage ± saffron ± tarragon ± thyme.

FRITTERS: balm, basil, caraway, chervil, chive, clary, comfrey, dill, fennel, mint, parsley, pineapple mint, rosemary, shallot, tarragon.

GIBLETS: chives, nutmeg.

GOULASH: caraway, chile, paprika.

GRAVY: caper.

GUACAMOLE: cayenne, chile, garlic, onion, tabasco.

GUMBO: cayenne, chile, garlic, onion, parsley, sassafras, tabasco, thyme.

HERB PEPPER: allspice, cinnamon, marjoram, pepper, rosemary, savory, thyme.

JAMS: angelica, balm.

JELLY: balm, geranium, lavender, mint, parsley, peppermint, sage, spearmint, tarragon, thyme.

LIPTAUER: caraway, mustard, onion, paprika.

MARINADES: basil, chile, fennel, garlic, juniper.

MINCEMEAT: allspice.

OMELET: balm, basil, celery, chervil, chive, clary, cumin, dill, garlic, marigold, marjoram, mustard, onion, paprika, parsley, sage, savory, tarragon, thyme, tumeric, watercress.

PASTA: basil.

PASTRY: basil, chives, lemonthyme, marjoram, mint, parsley, tarragon.

PICKLES: allspice, anise, bay, celeryseed, cinnamon, clove, dill, garlic, mace, mustard, nutmeg, tumeric.

PIE: allspice, anise, clove, fennel, ginger, nutmeg.

PIZZA: basil, chile, marjoram, oregano.

RAVIGOTE: burnet ± chervil ± chives ± tarragon.

SALAD: angelica, balm, basil, borage, burnet, celery, chervil, chile, chives, cicely, fennel, garlic, hyssop, lovage, marigold, marjoram, nasturtium, onion, parsley, pepper, poppyseed, rocket, rue, sorrel, tarragon, thyme, watercress.

SALAD DRESSINGS: balm, basil, bergamot, burnet, caper, chervil, chive, coriander, costmary, dill, garlic, geranium, ginger, horseradish, hyssop, lavender, lovage, marjoram, mint, mustardseed, nasturtium, poppyseed, rosemary, rue, sage, savory, tarragon, thyme, tumeric.

SALT SUBSTITUTE: Cloves, coriander, lemonpeel, paprika, pepper, rosmary, sassafras, tarragon, OR
aniseed, basil, garlic powder, lemon rind (or powdered lemon juice), oregano, OR
basil, celeryseed, cumin, lemonthyme, marjoram, sage, savory (Shimizu, 1982).

SCRAMBLED EGGS: basil, chile, chive, parsley, tansy, tarragon, thyme.

SLAW: caraway, mint.

SOUFFLE: chervil, chive, parsley, sage, savory, tarragon, thyme.

SOUP: balm, basil, bay, burnet, celery, celeryseed, chervil, chive, coriander, dill, fennel, hyssop, lovage, marigold, parsley, perilla, rosemary, saffron, savory, sesame, sorrel, tarragon, thyme.

SPAGHETTI SAUCE: basil, bay, chile, cumin, oregano, scallion.

SPICED SALT: bay, cayenne, clove, marjoram, nutmeg, pepper, thyme.

STEWS: allspice, celery, lovage, marigold, rosemary, rue.

SWEET AND SOUR: tamarind.

SYRUP: allspice, fenugreek.

TAHINI: garlic.

TINTS: annato, beet, calendula, cochineal, fennel, grape, marigold, paprika, saffron, tumeric.

TOFU: chile, cumin, onion.
VEGETABLE SOUP: basil.
VERMOUTH ADDITIVES: angelica, balm, caraway, celery, chamomile, cloves, dill, elecampane, fennel, gentian, ginger, nutmeg, peppermint, rosemary, sage, spearmint, staranise, sweetflag, thyme, wintergreen, wormwood, yarrow.
VINEGARS: basil, burnet, chervil, chile, dill, garlic, horseradish, lemonthyme, lovage, marjoram, mint, oregano, rosemary, sage, shallot, tarragon, thyme.
WINE FlAVOR: balm, basil, burnet, chamomile, cinnamon, cloves, galanga, ginger, hyssop, rosemary, sage, wintergreen, woodruff.

Recommended Culinary Combinations For Foods

Bay, chervil, parsley, thyme.
Basil, chive, parsley, savory.
Basil, parsley, savory.
Basil, chives, savory, thyme.
Basil, marjoram, parsley, thyme.
Burnet, chervil, chives, capers, garlic, tarragon, watercress.
Burnet, chervil, chives, tarragon.
Burnet, chives, parsley, shallot, tarragon.
Chives, fennel, savory, thyme (fish).
Chive, dill, garlic, tarragon.
Parsley, sage, rosemary, thyme.
Chervil, cicely, tarragon.
Chervil, garlic, parsley, shallot, watercress.
Chervil, chive, parsley, sage, savory, tarragon, thyme.
Basil, bay, cayenne, clove, garlic, mace, marjoram, nutmeg, pepper, savory, thyme.

Holly Shimizu (1982) of the National Herb Garden, USDA, has provided the following two lists.

STRENGTH OF CERTAIN HERBS

Strong or Dominant Flavors: Should be used with care; their flavors stand out; ca 1 teaspoon for 6 servings: bay, cardamon, curry, ginger, hot peppers, mustard, pepper (black), rosemary, sage.

Medium Flavors: A moderate amount of these is recommended; ca one to two teaspoons for 6 servings: basil, celery seed and leaves, coriander seed and leaves, cumin, dill, fennel, french tarragon, garlic, marjoram, mint, oregano, savory (winter and summer), thyme, tumeric.

Delicate Flavors: May be used in large quantities, combining well with most other herbs and spices: burnet, chervil, chives, parsley.

HERB BLENDS

Combining herbs for particular uses will save time and is especially good for specific foods. Blends can be added loosely or wrapped in cheese cloth and removed before serving. (The numbers refer to parts per blend.)

• *Egg Herbs:*	1 basil, 1 dill weed (leaves), 1 garlic, 1 parsley
• *Fish Herbs:*	1 basil, 1 bay leaf (crumbled,), 1 french tarragon, 1 lemon thyme, 1 parsley (options - fennel, sage, savory)
• *Poultry Herbs:*	1 lovage, 2 marjoram, 3 sage
• *Salad Herbs:*	1 basil, 1 lovage, 1 parsley, 1 french tarragon
• *Tomatoe Sauce Herbs:*	2 basil, 1 bay, 1 marjoram, 1 oregano, 1 parsley, (options - celery leaves, cloves)
• *Vegetable Herbs:*	1 basil, 1 parsley, 1 savory

- *Italian Blend:*   basil, marjoram, oregano, rosemary, sage, savory, thyme
- *Barbeque Blend:*   cumin, garlic, hot pepper, oregano

Guidelines

1. Herbs should not flavor more than two courses in a given meal.
2. Herbs should be ancillary will-o-the-wisps, not dominant elements in the course.
3. Don't judge the strength of the herb's contribution until the dish has been cooked for some time.
4. Of the strong herbs, like lovage, rosemary, sage, and tarragon, only one should be used in quantity.
5. Basil, parsley, and savory can be combined with the stronger herbs like lovage.
6. Chervil, chives, and sweet marjoram are good mixers.
7. Fresh is better than frozen is better than shade-dried is better than sun-dried.
8. Some must be used fresh: anise, balm, borage, burnet, chervil, coriander, geranium, lemonverbena, nasturtium, savory, sorrel, violet.
9. Others are best fresh, yet usable prepared: basil, chives, marjoram, rosemary, tarragon, thyme.
10. Herbs are usually better if dried, and stored, in the absence of sunlight.
11. Ulcer dieters can eat several mild herbs, but no pepper and hot spices. Basil, chervil, marjoram, and savory can partially replace salt and pepper for seasoning.
12. Too many herbs spoil the soup!
13. Basil, fennel, mint, parsley, rosemary and tarragon are frequently used alone while chervil, savory and thyme are more often used in combination (Page and Stearn, 1979).

CHAPTER 3

Herbal Teas

December 16 is the anniversary of the Boston Tea Party. In 1773, citizens disguised as Indians dumped 340 chests of tea into the sea, one of the triggers for our American Revolution. Tea is still important to Americans, but I sense a second revolution, another tea party in the making.

Camellia sinensis is the species from which most North American teas are made. Tea made from *Camellia* is certainly not on the decline! We imported 54 million dollars worth in 1970, $65 million in 1973, and nearly $130 million in 1982. Similar figures for coffee are in the billions. Leaves of both these species contain the stimulant caffeine, a relatively mild toxin (LD50 orally in rats, 192 mg per kg bodyweight). The revolution I sense is that a lot of new-fashioned people are experimenting with old-fashioned herbal teas. Strange plants, some of them dangerous, are sneaking into the tea bags.

Julia Morton recently published a paper entitled "is there a safer tea?" She alluded to dozens of herbal tea mixtures, pointing out that some of them were really quite dangerous. A magazine my family takes has advertisements for acerola, alfalfa, almond, aloe, blueberry, buckwheat, camomile, carob, chickpea, comfrey, fenugreek, garlic, lemon, papaya, parsley, peppermint, pumpkin, rosehip, sassafras, and wintergreen. Several of these, like sassafras, probably contain carcinogens. But several if not all of our foods probably contain carcinogens. Safrole, e.g., is the carcinogen in sassafras, once found in certain types of rootbeers. It has been taken out of the rootbeers, but it's still in the sassafras tea my 101-year-old grandmother offered me. Small amounts occur in several spices in your spice rack. You'll also find it in toothache packs, with hops and chile, prescribed by modern dentists.

Coffee and tea both contain caffeine, also found in cocoa, cola, guarana, and mate (Paraguay Tea). Like many teas, they may also contain tannic acid. The astringency of the tannin is desired by some tea devotees. But some writers suggest that consumption of hot herbal teas and their tannin correlates with high incidence of cancer. Both caffeine and tannic acid are listed in Toxic Substances 1974, caffeine even being teratogenic in mice. Total acceptable daily intake of tannin for man is 560 mg, but heavy coffee, tea, or cocoa drinkers may get twice this. High tannin is especially harmful with young people on low protein diets. In other countries, exhaustive decoction of 100 g tea/day contribute close to 30 g of polyphenols, leading to toxicity.

It's sort of depressing; each week a finger is pointed at something. I was once a three-pack-a-day smoker but I quit. Then I started drinking more coffee and tea. My doctor then told me to cut down on caffeine, so I sampled some of the herbal teas. I found various mixtures of mints, especially lemon-scented mints, geraniums and verbenas, with raspberry, strawberry or sassafras leaves quite satisfactory, especially when sweetened. But then I compared the list of chemicals found in my herbal teas with Toxic Substances 1974. I've been drinking minute quantities of anethole, borneol, carvacrol, carvone, citral, citronellol, coriandrol, cymene, esdragol, eucalyptol, eugenol. fenchone, linalool, pinene, safrol, and thymol. All are listed in TOXIC SUBSTANCES 1974 but none are as toxic as caffeine, based on their LD50 (Duke, 1977). After 10 years of reading about herbal teas, I dared to venture my appraisal of 365 of them, as to whether I think they are safer or more dangerous than coffee. The manuscript was submitted to CRC Press in 1983 with the title An Herb A Day ... Borderline Herbs.

Not all teas are pleasant to the taste. Many are popular because of their reputed medicinal virtues. Although you can buy ginseng tea bags, I doubt that many people are drinking ginseng for its flavour. Perhaps the old spouses' tale, that it clarifies the mind and rejuvenates the body drives, increases its popularity. A recent study suggests that there is scientific evidence to support that old spouses' tale. Six saponins (many sex hormones are saponins) were isolated by Japanese scientists studying Korean ginseng. Extracts stimulated the production of albumin and globulin in mice.

Parsley, sage, rosemary, and thyme are all used in herbal teas. Rosemary and thyme are frequent in aromatic teas, parsley and sage perhaps less frequent. I boiled all four in different containers and tried various mixtures. I recommend more rosemary and thyme, less sage and parsley, the latter two for pungent vitamin-rich body, the former for aroma. To each his own; some people fashion their teas and/or their salads from pure parsley. I prefer rosemary and thyme for an aromatic tea.

But the parsley I'll save for my broth or my "vegetable juice cocktail." In Ag Handbook No. 8, you'll find ten weedy vegetables, amaranth, chicory, dandelion, dock, gardencress, lamb'squarters, mustard, poke, purslane, and watercress. Five ounces of these weeds (dried) contain more than your RDA in protein, calcium, phosphorus, iron, vitamin A, vitamin C, thiamin, riboflavin, and niacin (see Chapter 7). I cooked all these, adding salt and pepper, basil, dill, parsley, thyme, and wild onion, and made a satisfactory broth or "potlikker." Ounce for ounce the dried weed leaves are richer in all these nutrients except niacin and thiamin than one "fortified cereal." How much nutrient was in the "potlikker" and how much was in the dregs, I don't know. Nutritional analyses of some of these wild greens, all more than 80% water when gathered, are presented in Table 7. The average has been converted to a zero moisture basis there, to compare with the herbs, analyses of which are presented in Table 11).

Table 7. Analysis of Shoots and Greens

Shoots

Food Item	H₂O g	Cal	Prot g	Fat g	Total Carb g	Fiber g	Ash g	Ca g
Asparagus	91.7	26	2.5	.2	5.0	.7	.6	22
Bamboo	91.0	27	2.6	.3	5.2	.7	.9	13
Mungbean	88.8	35	3.8	.2	6.6	.7	.6	19
Pokeweed	91.6	23	2.6	.4	3.7	—	1.7	53
Soybean	86.3	46	6.2	1.4	5.3	.8	.8	48
Average (APB)	89.9	31	3.5	.5	5.2	.6	.9	31
(ZMB)	0	307	34.6	5.0	51.5	5.9	8.9	307

Greens (wild)

Food Item	H₂O g	Cal	Prot g	Fat g	Total Carb g	Fiber g	Ash g	Ca mg
Amaranth	86.9	36	3.5	.5	6.5	1.3	2.6	267
Dandelion	85.6	45	2.7	.7	9.2	1.6	1.8	187
Dock	90.9	28	2.1	.3	5.6	.8	1.1	66
Lambsquarters	84.3	43	4.2	.8	7.3	2.1	3.4	309
Purslane	92.5	21	1.7	.4	3.8	.9	1.6	103
Average (APB)	88	35	2.84	.54	6.48	1.34	2.1	186.4
(ZMB)	0	292	23.7	4.5	54.0	11.2	17.5	1,553

Source: Duke and Atchley, 1984

P mg	Fe mg	Na mg	K mg	B-Car ug	Thia mg	Rib mg	Nia mg	Vit C mg
62	1.0	2	278	540	.18	.20	1.5	33
59	.5	—	533	12	.15	.07	.6	4
64	1.3	5	223	12	.13	.13	.8	19
44	1.7	—	—	5,220	.08	.33	1.2	136
67	1.0	—	—	48	.23	.20	.8	13
59	1.1	4	345	1,166	.15	.19	1.0	41
584	10.9	40	3,416	11,543	1.48	1.88	9.9	405

P mg	Fe mg	Na mg	K mg	B-Car ug	Thia mg	Rib mg	Nia mg	Vit C mg
67	3.9	—	411	3,660	.08	.16	1.4	80
66	3.1	76	397	8,400	.19	.26	—	35
41	1.6	5	338	7,740	.09	.22	.5	119
72	1.2	—	—	6,960	.16	.44	1.2	80
39	3.5	—	—	1,500	.03	.10	.5	25
57	2.66	40.5	382	5,652	.11	.24	.9	68
475	22.2	337	3,183	47,433	.9	2.0	7.5	567

Pokeberry is poisonous, yet it's one of the weeds listed in Ag Handbook No. 8. Properly cleaned and cooked it is safe, but how many herbal tea makers can identify all the herbs, let alone pokeberry?

Herbal teas can affect the body. Ailments treated by parsley, sage, rosemary, and/or thyme, if you can believe all the herbals, include: acne, ague, alcoholism, alopecia, amenorrhea, anemia, asthma, baddreams, blepharitis, bronchitis, bruises, cancer, cataracts, cattarh, chilblains, chlorosis, cirrhosis, cold, colic, conjunctivitis, coughs, cramps, cuts, dandruff, debility, delirium, depression, diabetes, diarrhea, dropsy, dysentery, dysmenorrhea, dyspepsia, earache, eczema, enteritis, epilepsy, eyes, fever, flatulence, gallstone, gastritis, gingivitis, gout, gravel, grief, halitosis, hangover, headache, headcold, heart, hemorrhage, hookworm, hysteria, impotency, indigestion, insect bites, insomnia, intoxication, jaundice, kidney, laryngitis, leprosy, lethargy, leucorrhea, liver, lungs, malaria, measles, memory, menstruation, mycosis, nerves, neuralgia, otitis, palpitation, palsy, parasites, phthisis, piles, plague, psoriasis, quinsy, rabies, rheumatism, ringworm, salivation, scarlet fever, sciatica, scrofula, scurvy, shakes, short breath, sinusitis, snakebite, sorethroat, spasm, spleen, stitch, stomach, stones, teeth, tonsilitis, trembling, tuberculosis, tumors, typhoid, ulcers, uterus, vertigo, vision, warts, whooping cough, worms, and wounds.

I don't believe that parsley, sage, rosemary, and thyme can cure all these ails. But someone who ingests these herbs ingests several "toxins," among them acetaldehyde, alcohol, apiol, borneol, carvacrol, cineole, myristicin, pinene, tannin, thujone, and thymol. According to the 8th edition of Merck's Index, these "toxic" chemicals are used medicinally in the treatment of bronchitis, diarrhea, dysmenorrhea, fever, hysteria, insomnia, laryngitis, mycoses, pharyngitis, rhinitis, and worms. This might suggest that there is some scientific rational behind herbal folk medicines. All four of the herbs are found in Hartwell's "Plants Used Against Cancer." If only 25% of the folk cures showed some scientific rational behind their folk usages, odds are that one plant of the four would be effective. I've emphasized parsley, sage, rosemary, and thyme, mainly because of my fondness for an old English Folk Song, Scarborough Fair. Other tea ingredients are mentioned in Chapters 4-9. Sage tea might be recommended for colds on the basis of folklore. Which makes more sense, a witch doctor prescribing antibiotics for a suspected cold, or an old spouse prescribing sage tea, more favored than our tea by the Chinese, for a cold? What is a cold? What is the cure?

A lemon flavor for our tea may be provided by leaves of any of several herbs, some not even closely related; lemongrass of the grass family, lemonwood tree of the Pittosporum family, lemon verbena from the verbena family, lemon geranium of the geranium family, lemon eucalyptus of the myrtle family, trifoliolate orange of the orange family, and my favorites, lemon balm, lemon basil, and lemon thyme of the mint family. Sweeteners may be provided by the tropical miracle fruit, *Synsepalum dulcificum, Dioscoreophyllum cumminsii,* or *Thaumatococcus daniellii.* Roots of licorice and several other legumes, some of them poisonous, contain the substance glycyrrhizin, and are sometimes used as sweeteners. Paraguay's *Stevia rebaudiana* contains a substance 300 times sweeter than sugar, but it's also reported to be contraceptive. More interesting to me, perhaps because it's a weed in my back yard, is *Perilla frutescens.* It also contains an estrogen hormone, but it contains the substance perillaldehyde, starter for a substance 2000 times sweeter than sugar. A little *Perilla* makes a good mixer in aromatic tea blends, but so far, I've not discovered how to release the sweetening agent. Once I learn that I can grow all my tea ingredients in my kitchen window. I don't use cream, but those who do are doing themselves one favor. The cream is said to inactivate the "toxin" tannin.

In May, 1984, in the remote village of Gandoca, Costa Rica, with no electricity or plumbing, five of us were reduced to drinking a strong tea of ginger and lemongrass after dinner of rice and beans. The sweetened ginger and lemongrass tea was as good as any herb tea I've had. But sitting around breakfast, also rice and beans, the next morning, we realized that 60% of us had experienced mild hallucinations during the night. Was it the tea?

75

CHAPTER 4

Herbal Liqueurs

The man or woman fortunate enough to have a sunny window in his home or apartment can get many hours satisfaction growing a window herb garden. Most of the herbs used in making herbal teas are equally good or better bases for making homemade liqueurs. The cheapest of ethyl alcohol, white wine, eau de vie (a colorless fruit brandy), gin, or vodka can be made to simulate some of the most expensive liqueurs. The herbs may be steeped directly in the alcohol, or boiled down to a water concentrate, with or without added sugar. If you like the aroma of the herb or a combination of herbs, and you like liqueurs, chances are you can make a liqueur as good as some of your favorite brands, maybe better, surely cheaper. First try a tea made by steeping or boiling the herb or combination of herbs. If you like the tea, concentrate it by boiling or evaporation, and add alcohol and sugar to taste, first on a small scale, so you won't waste that expensive alcohol. Once you have savored your sample, then you may feel it worth making a bigger batch. With hundreds of herbs to choose from, simple or in combinations, there is nearly an infinite array of possibilities for your own secret formulas. Following are brief writeups on several herbs that show promise in liqueur preparation. I have grown almost all of them successfully and tried most of them in homemade liqueurs. I truthfully have enjoyed many of my concoctions at $2.00 per pint, much more than many of those $10.00 per pint commercial liqueurs, most of which are based on closely guarded secret recipes, containing dozens or even more than 100 herbal ingredients. I find my homemade mixes of the handiest dozen herbs more attractive than the simpler one- or two-herb liqueurs.

Many sophisticated Americans would view their ancestors as naive for drinking teas made of herbs that modern Americans rarely encounter. Still if they are liqueur connoisseurs they may be consuming a wider variety of herbs than their ancestors. It's all but impossible to find out the secret recipes of many of these liqueurs, but some are said to contain 100 or more herbs. Many modern Americans couldn't even name 100 herbs. Leung (1980) lists more than 100 GRAS herbs and or species used in alcoholic beverages. See Table 8. See also the con cluding chapter Herbs and Man. For those Americans who like liqueurs but know little about herbs, I've pulled this chapter together. I've made a few liqueurs at home very cheaply that I prefer to those high-priced syrups my more sophisticated friends feign to prefer.

Liqueurs can be based on bourbon, brandy, Irish, rum, scotch, or straight whiskey. I have two big objections to using these. They generally have more overpowering flavors and prices higher than gin, vodka and white wine. I'd hate to ruin even a good gin or vodka with a bad herbal combination. So unlike most American experimenters with liqueurs, I do not first steep an herb in the expensive alcohol. I save a lot of herbs and liquors that way. I first try the combination of herbs in a tea or broth. I think of it as a tea if I want to sweeten it, broth if I want to salt it. (Pink clover, chicory flower, and wild garlic make a good broth). The tea combinations that I like, I boil down to concentrated sugar infusions making a syrup to add to vodka to make the liqueurs. The broth combinations that I like, I concentrate by boiling, with or without adding sugar.

These concentrated broths I add to gin to make more bitter liqueurs. So my bitters go with gins, my sweets with vodkas. When I want to get fancy, I add glycerine to thicken and artificial or natural dyes to tint. Once satisfied with a combination, I'm not reluctant to steep the herbs in the appropriate alcohol base. As a rule one does get better and more interesting boquets from steeping, since many of the more elusive oils lost in boiling are retained in steeping.

Without bragging about my liqueurs, I relate one true anecdote. I once had six homemade liqueurs, which I colored and thickened with glycerine to match the one "storebought" liqueur which will remain anonymous. The seven taste panelists, from seven walks of life, not knowing how many were homemade, how many storebought, unanimously singled out the solitary storebought ($9.00 a fifth) as bad! None of the others, all homemade, were rated that low.

A cup of honey is preferred by connoisseurs to a cup of sugar in most liqueur recipes. But until I'm producing my own honey, I'll experiment with sugar. In the herb *Perilla* there's an ingredient which with a slight chemical manipulation is converted into a compound 2000 times sweeter than sugar. I've learned to grow the perilla, but not to convert the sweetener compound. There's an ingredient in *Stevia,* a subtropical species, that's 300 times sweeter than sugar. One leaf, of a batch I have, collected in 1945, will still sweeten a cup of coffee. My sugar-leaf *Hydrangea* died, after sweetening many of my herbal teas and liqueurs. Now I'm experimenting with my "mayapple brown sugar", made by drying mayapple sauce in the sunshine. One of my favorite teas for 1983 I called the "Red Ringer" in allusion to the resemblance of my tea to the famed Red Zinger. "Red Ringer" is tinted with wild bergamot, flavored with wild ginger and lemon balm, and sweetened with mayapple pulp. With vodka it makes a great liqueur. Since mayapple, an old Indian cancer remedy, according to Hartwell (1967-71), has now turned out to be the second most important plant in the world in the fight against cancer, I'm particularly prone to experiment with it. The Indians also said it would work against insects so I induced the USDA to check that out. The Indians were right again. The root extract strongly deters feeding activities of potato beetles.

Alcohol kills thousands of Americans each year, and alcohol nearly killed me, both chronicly and acutely. I do not wish to encourage the consumption of ethanol, the killer, any more than I want to promote the consumption of herbs, the healers.[1] Even herbs claim a few lives each year, especially those abused herbs like coca, datura, marihuana and poppy. If you drink too much, you might taper off a bit if you resort to making a new herbal liqueur between each drink. At least that will be a constructive and interesting way to slow yourself down. If you insist on taking ethanol, with or without herbs, DO SO AT YOUR OWN RISK. I am relating what I have tried myself or read, but I am not encouraging herbal self medication. Only your doctor can legally prescribe medicine to you. There are increasing numbers of MD's today, perhaps cognizant of the increasing numbers of iatrogenic diseases, who are increasingly prescribing natural herbs. I am not! Just because I have tried and enjoyed or benefitted from an herbal broth, herbal liqueur, herbal medicine or herbal tea, does not mean that you should. To avoid the possibility of a lawsuit from someone who thinks I'm promoting herbs, I restate here that I AM NOT RECOMMENDING HERBS FOR ANYTHING. I am recounting my own experiences and those of others I've read about for your information. I candidly believe that many GRAS herbs are less liable to hurt you than alcohol, caffeine, and/or tobacco, and I'm certainly not recommending those either.

Of the Herbalist's Dozen (Chapter One), at least basil, caraway, dill, fennel, marjoram, oregano, sage, tarragon and thyme have shown up in herbal liqueurs. Short accounts for some other herbs useful in liqueurs follow.

---

[1] Some respected authorities estimate that more than half our medicines are or were at one time derived from plants or other organisms.

ABSINTH
(Artemisia absinthium L.)

CULTURE: Seeds sown in autumn. Cuttings or root divisions of this bitter *perennial* may be spaced 2 feet apart in a sunny or partially shaded well-drained loam or clay loam. Once started, no further attention is needed, save for a little weeding. Yields of 500 lb/a dry inflorescence are possible.

USES: With the possible exception of rue, wormwood is one of the most bitter herbs. Vermifuge, moth and insect repellent, it is said to resist putrefaction. It has been used by brewers instead of hops. Recently new evidence suggests that absinth might work on the brain in the same way that THC, the marihuana magic, works. Small wonder that mixed with marigold, marjoram and thyme, absinth found its way into love potions. Absinth, bearing also the name "Wermuth" (preserver of the mind) was first known as a nervine and mental restorative. Inferior absinth was often adulterated with copper, which gave it a characteristic green color. Hemingway regarded absinth as an aphrodisiac. Like many other substances, absinth *may* be good in small doses, bad in big doses. But absinth was outlawed as a beverage, for better or worse. Commerical wormwood-free substitutes appeared, among them Abisante, Abson, Herbsaint, Nistra, Oxygene (Ojen), and Pernod. Herbalists who wish to live dangerously can grow and make their own absinth, remembering that wormwood is very bitter. Wormwood and anise are basic; I prefer more anise, less wormwood. Optionals include almond, angelica, balm, cardamon, cinnamon, clove, coriander, elecampane, fennel, ginger, lemon, marjoram, orange, orris, and peppermint. Equal parts of aniseed, coriander, fennel, and marjoram, with a trace of wormwood, and some ground angelica, in sugared vodka for a few days should do the trick. The Rodale Herb Book (Hylton, 1974) suggests a mixture of common and Roman wormwood, lavender and lemon verbena for a homemade herbal liqueur. Taken to excess, the drug absinthium produces giddiness and attacks of epileptiform convulsions. The chalky white skin in Toulouse-Lautrec's the Absinthe Drinkers is said to be due to excess wormwood. Absorbine is said to be based on absinth.

FOLKLORE: Absinth is regarded as anthelmintic, antiseptic, antispasmodic, aperient, aphrodisiac, cardiac, carminative, cholagogue, febrifuge, stimulant, stomachic, and tonic. The bitter tea, recommended for women with labor or menstrual pains, might also be useful in cachexia, colic, diarrhea, dropsy, dysentery, dyspepsia, fever, gallbladder ailments, gravel, heartburn, liver ailments, malaria, migrain, pyorrhea, rheumatism, sores, stones, and other urinary ailments, as well as worms. Wormwood oil allays pain applied externally in arthritis, neuralgia, and rheumatism.

A wormwood cordial, an ounce of wormwood, steeped for 6 weeks in a pint of brandy, ruins a good pint of brandy from the culinary point of view but is supposed to aid gout and gravel. I can recommend my Hot Gold Martini in which absinth or Roman Wormwood *(Artemisia pontica)* is substituted for vermouth, lemon balm for the twist of lemon, marigold petals for the gold, and gin for the glow. If you like a wet martini, add a dash of sugar and a cube of ice.

ANGELICA
(Angelica archangelica L.)

CULTURE: Start with seed less than a month old, or buy three *biennial* or *perennial* plants, one for root, one for seed, and one for a rainy day. Plant in a good loam getting noonday shade. Pinch off the flower buds to keep the plant perennial. It dies when it goes to seed. Yields of 1,500 lb/a dry root are possible.

USES: Roots and seeds are used to flavour bitters, liqueurs and vermouths. Oil of Angleica Root commands more than $100. per lb., probably because it is the most important herb in several liqueurs, Benedictine, Chartreuse, Drambuie, Galliano and Strega. To simulate yellow Chartreuse, take 2 spoons of chopped angelica root, 4 of basil, 4 of hyssop, a short cinnamon stick, a dash of mace, adding marigold or saffron for color to a quart of sugared vodka. Balm can be substituted for the basil. Drambuie is simulated by steeping a small angelica root in Scotch Whiskey overnight, adding honey to taste. To simulate Galliano, leave out the balm in the "Chartreuse," adding instead a clove and dash of nutmeg to the quart of sugared vodka. To simulate Strega, use 3 spoons of chopped angelica, 4 of aniseed, 10 crushed cardamons, a clove, and a dash of mace and nutmeg to the quart of vodka. In spite of its marvelous adaptability to liqueurs, and its being a flavoring with juniper in British gins, an old wive's tale persists that if candied angelica stems are taken in quantity, they cause an aversion to alcoholic beverages. It is even rumored to sedate the lusts of young people. Seeds find their way into vermouth and the leaves are used in making bitters (one oz. angelica, one oz. holy thistle, ½ oz. dry hops, infused in 3 pints boiling water). Rhine wines may owe some of their flavour to angelica. French Absinthe contains both angelica and absinth. Leaves are said to be smoked by Laplanders. Used in cosmetics, said to quiet or sooth the nerves of the skin (Leung, 1980).

FOLK MEDICINE: In herbal teas, it makes a good digestive, whose beneficial effects may persist for days. It has been used in folk remedies for alcoholism, anemia, arteriosclerosis, bronchitis, chilblains, colds, cough, cramps, eyes, fever, flu, gout, headache, indigestion, leucorrhea, lungs, malaria, plague, pleurisy, rheumatism, scabies, scrofula, stomach, toothache, typhoid, typhus and ulcers.

ANISE
(Pimpinella anisum L.)

CULTURE: Sow seeds of this *annual* closely in cool weather, praying for hot weather after germination. Sow in a sunny spot with light soil. Thin to one foot apart. Keep weeds down. Allow four months until harvest. Don't pick the seed until they are ripe, and don't store until they are dry and insect free. Planted at 5-10 lbs/a, plants may yield 500-2000 lbs/a (Duke, 1978; Rosengarten, 1978).

USES: The licorice-flavored anise is used more and more as a substitute for licorice. Still aniseed deserve their own role in seasoning, confectionary, cookies, soups and stews, more importantly in liqueurs. One good anise liqueur is made by stirring 6 spoons crushed aniseed in a quart of brandy. If you have as much trouble as I do growing anise, you can substitute fennel seed. Anisette combines equal parts (2 spoons each) of rather equal flavored anise, coriander, and fennel seed in a quart of sugared vodka. Anisette is said to be good for bronchitis. A pleasant way to take your vitamins would be with Farrel's Rose-Hip and Anise Liqueur, which can be simulated by boiling aniseed and rosehips down to a syrup in sugar water to add to vodka. For those not in a hurry, steeping in the vodka is favored over boiling, but weekend liqueur connosieurs can achieve quicker results using the herbal syrup approach. The Rodale Herb Book (Hylton, 1979) suggests anise or fennel seed with orange or lemon peel as the basis for one liqueur. I've enjoyed the combination of all four with gin in the liqueur I call Citrus Fanisey. Allasch is a Latvian Kummel with anise, almond and caraway. Anesone is an anise-flavored cordial, sweeter and stronger than anisette. Ojen is a Spanish liqueur, high in alcohol and anise. Ouzo is a Greek anise liqueur. A French anise-based liqueur is called Pastis. Tres Castillos is a Puerto Rican anise-flavored liqueur. Anise makes a nice ancillary ingredient to other more delicate liqueurs but should be used cautiously. Aniseed in warm milk promotes milk production as well as sleep in nursing mothers, and can be used as a collyrium. In old days aniseed were valued against the evil eye and the bad breath. If you grow a good quantity of seed, they are useful for refreshing the breath. Some people smoke the seed to clear the throat. Teas and salads are embellished by the addition of small quantities of leaves. Harrop (1977) offers a recipe for aniseed and sesame biscuits.

FOLK MEDICINE: Regarded as antiseptic, antispasmodic, aperient, aphrodisiac, aromatic, carminative, digestive, expectorant, galactagogue, pectoral, stimulant, stomachic and tonic, anise is supposed to help asthma, bronchitis, coughs, dropsy, dysmenorrhea, epilepsy, gallbladder, halitosis, indigestion, insomnia, lice, migrain, nausea, neuralgia, scabies, stomach, and stones. Placed under the pillow, anise is supposed to ward off bad dreams.

AVENS
(Geum urbanum L.)

CULTURE: If you can find them, seeds of this *perennial* will suffice for a start, but most of the plants are collected from the wild.

USES: The root is used as a spice, and could be substituted for clove or cinnamon in various recipes. The volatile oil is sometimes used as a substitute for cloves. Placed among linens, the plant is supposed to keep moths away in addition to imparting a pleasant aroma. Augsburg Ale and other beers are said to be improved by Avens which prevents them from turning sour. Mixed with orange peel, it is added to wines to make a vermouth-like cordial. The fresh root imparts a clove flavour to liqueurs. One old recipe calls for avens, angelica and tormentil with raisins in brandy or in wine.

FOLK MEDICINE: Antiseptic, aromatic, astringent, febrifuge, stomachic, styptic, sudorific, and tonic, the roots are used for ague, catarrh, chills, colds, debility, diarrhea, dysentery, gastroenteritis, headache, hemorrhage, leucorrhea, liver, malaria, sore throat. Chewing on the root is supposed to prevent cavities and bad breath.

BERGAMOT
(Monarda didyma L.)

CULTURE: Setting few seed, this handsome *perennial* is easily propagated by clump divisions and readily takes root in the shaded moist soils which favor it. The bright red flowers, more than the leaves, are harvested for tisanes, cordials, and potpourris. It should be subdivided annually or the plant deteriorates. Harvested leaves and flowers should be dried in a a warm but shaded well-ventilated spot. My leaves are almost always coated with a powdery white, which I assume is a mildew. Yields of 1,000-1,500 lb/a leaf seem feasible.

USES: Attracting bees and hummingbirds, this handsome plant is worth adding to the garden solely as an ornamental. An infusion of the plant, heavily laced with a nice fragrance, was once commonly used for tea in the United States. The plant is said to yield thymol, a powerful antiseptic, deodorant, disinfectant, and meat preservative. In higher animals, thymol acts as a local irritant and anaesthetic to the skin and mucous membranes. An oily solution of thymol is applied to the respiratory passages in stuffed noses, and it is inhaled for laryngitis, bronchitis, and whooping cough. Internally it is used, somewhat dangerously, as a vermifuge and in treating diabetes. It is applied as an antiseptic lotion for burns, chilblains, eczema, parasitic skin infections, and psoriasis. Burning the plant might release enough thymol to serve as an insect repellent. Bergamot may be substituted for horsemint *(Monarda punctata)* as a carminative, diaphoretic, diuretic, emmenagogue, rubefacient, stimulant, and vesicant to rub on rheumatic sore spots. Tea of bergamot alone has a musty thymol taste to me, and reminds me somewhat of the aroma of rabbit tobacco. The tea might be a useful aproach to treating those ailments treated with thymol. The tea is recommended for headaches, and a few leaves might be added to bath water for fragrance. I recommend the addition of lemon balm in either case.

It took the American Revolution to induce patriotic Americans to use the tea used by the Oswego Indians while they boycotted the British tea at the Boston Tea Party. Teas made from the leaves and flowers have a soothing and sedative effect. Leaves are used to flavor fruit cups, jellies and salads. Fresh or dried leaves and/or flowers spike the flavours of wines and cocktails. Alone, but better in combination with rosemary, or thyme, this makes a good Bergamot Liqueur with sugared vodka. Boiled with catnip and sugar, it makes an interesting liqueur with gin. The petals from one plant are sufficient to tint a tea or liqueur bright red.

FOLK MEDICINE: Carminative, diaphoretic, diuretic, emmenagogue, rubefacient, soothing, stimulant, bergamot is used, usually as a tisane, to alleviate cholera, colds, flatulence, nausea, rheumatism, sorethroat, and urinary disorders.

BORAGE
(Borago officinalis L.)

CULTURE: Sown and self-sowing by seed, this colorful *annual* thrives in sandy loams, well drained and lit. Once started and weeded, it tends to shade out the competition. Leaves and flowers should not be harvested until the dew has evaporated. It adds cheer, potted in a sunny window.

USES: As with beebalm, both flowers and leaves make welcome additions to wines, cordials, tankards of ale, and the blue flowers are useful for imparting decorative colors to Borage Liqueurs. According to older herbals, borage flowers and leaves in wine make man and woman glad and merry, and drive away all sadness, dullness, and melancholy. According to Dioscorides, borage was the famous Nepenthe of Homer, which when drunk steeped in wine, brought absolute forgetfulness.

The leaves may be substituted for cucumber in salads and may be boiled as a spinach-like vegetable. Flowers may be candied by dipping them in egg-white, then sugar. Tisanes made from the leaves, topped off with flowers, soothe throat and speed up circulation. Borage tea, heavy with lemon and/or balm is recommended for colds.

FOLK MEDICINE: Alexeteric, aperient, calmative, demulcent, diuretic, emollient, febrifuge, galactagogue, pectoral, and tonic, borage has been recommended for cold, fever, itch, jaundice, lungs, melancholy, peritonitis, pleurisy, ringworm, swelling, and ulcers.

BURNET
(Sanguisorba officinalis L.)

CULTURE: Sown early in a light soil, this bushy *perennial* can be propagated by root division in autumn. Plants should be spaced no closer than 9 inches. Flowers should be picked off;

otherwise the plant may self seed. Once established, the plant needs little attention. It is easily grown in a window pot, but why bother?

USES: Formerly regarded for the green fodder it produces in a mild winter, now used more for a cucumber-flavored addition to your garden salad after frost has nipped your cucumber and borage. It has been much used in wines and in salads as "pimpernella", essential in the best French and Italian salads. Steeped in wine, burnet was said in the olden herbals to comfort and rejoice the heart, and ease palpitations thereof. If you can scrape up a gallon of the reddish flower heads and 2½ lbs. sugar, you can have a wine after 3 days aerobic fermentation, with Chablis yeast, some citric acid, and tannin (Zanelli, 1972).

Leaves are added to beverages, cheeses, garnishes, herb butters, salads, soups, and vinegars. Fresh leaves are far superior to dried leaves. It is recommended with asparagus, bean soup, celery, mushroom soups and with salads dressed with French dressing or mayonnaise.

FOLK MEDICINE: Astringent, diaphoretic, hemostatic, and tonic, burnet has been recommended for diarrhea, dysentery, gout, heart, hemorrhage, leucorrhea, phlebitis, plague, rheumatism, varicose veins, and wounds.

CARDAMON
(Elettaria cardamomum (L.) Maton)

CULTURE: Unless you have a tropical rainforest in your backyard, you may have trouble with this *perennial*. Frost-sensitive, this can only be grown in a humid hot house up north. So far, I have been unable to produce any cardamons in my small greenhouse. Simulation of a shaded well-drained tropical forest soil is necessary. Seeds are usually sown immediately or stratified in shade for 2-9 days before sowing. Most seeds germinate in 6-8 weeks but some stragglers may come up only after a year. Yields run only 100-500 lbs dried capsule/acre. "Cardamon" geranium may suggest the aroma of cardamon but I don't know that it can substitute for cardamon.

USES: According to The Wealth of India (CSIR 1948-76), cardamon serves as a masticatory, medicine and spice. As a spice, it is used for flavoring beverages, like coffee, bread, cakes, curries, even liqueurs, and pickles and sausages. It is also used in perfumery. Some chew the cardamon hoping to mask the smell of alcohol. In 1983, the FDA challenged the safety and efficacy of a mixture of charcoal and fructose which was being promoted to mask drinkers' breath against the "breathalyzers". Perhaps the addition of cardamon, coriander, and parsley ground up with the charcoal would aid the masking process.

FOLK MEDICINE: Medicinally it is considered antispasmodic, aromatic, carminative, stomachic and stimulant. Seeds are chewed for indigestion and to mask unpleasant odors.

CATNIP
(Nepeta cataria L.)

CULTURE: Like many other mints, this hardy *perennial* can be started easily from seed, cuttings, or root division. Pinching back the flowers induces a bushier habit and increases the frost hardiness of the plant. The hairiness of the leaves impart some drought tolerance to the plant. And it will grow in partial shade. I rank it among the brown-thumb specials along with several other perennial herbs, which require almost no attention once started: balm, catnip, fennel, marjoram, mint (applemint, peppermint, spearmint), rue, sassafras, southernwood, tansy, thyme, and wormwood. Shade leaves though larger are less pungent as a rule. Leaves should be air-dried in the shade, and steeped, not boiled. Tolerating limestone well, catnip makes an interesting border, especially combined with hyssop. Foliage yields of 1,000-3,000 lbs/acre are possible.

USES: Although it turns felines and some hippies on, catnip is said to repel insects and mice. I'm sorry to say it did not repel the greenhouse mouse who ate my best ginseng root. The tisane is said to be good for inducing sleep, free of nightmares. The leaves are also used in French cuisine. In the Middle Ages, they were used to flavour salads. Catnip leaves, dipped in egg and lemon juice, then glazed, make an interesting dessert. I recommend as an eyeopener my recipe for Golden Catnip Tea. For each cup of boiling water, add 6 medium catnip leaves, the center of a small marigold (Tagetes) flower, and one spoon of sugar. Boiled down, such a syrup would contribute to a reasonable substitute for Chartreuse by adding angelica, balm, and hyssop or rosemary. Catnip syrup alone makes a good liqueur with gin while it seems better mixed with balm, mint, rosemary, and/or thyme for a vodka-based liqueur. Catnip and fennel teas or liqueurs might be useful in mild cases of colic.

FOLK MEDICINE: Regarded as anodyne, antispasmodic, aromatic, carminative, diaphoretic, emmenagogue, catnip has been recommended for bronchitis, cold, colic, diarrhea, erotomania, fever, headache, hives, hysteria, impotency, indigestion, inflammation, insanity, insomnia, labor, measles, nervousness, neuralgia, nightmares, piles, smallpox, stomach, tuberculosis, uterus, and vertigo. In old herbals it is suggested that chewing on the roots will make a gentle person cruel.

CELERY
(Apium graveolens L.)

CULTURE: Slow starting and slow growing, this well-known *biennial* vegetable is not so well known as an herb. Start seeds indoors, quite early, for transplant to the field. The slowly germinating seedlings require hardening for frost and drought. Originally the plant was rather tolerant of saline marshy soils. It is difficult to grow the vegetable, easier to grow the herbage. Leaves, as opposed to celery stalks, can be harvested anytime, and used fresh if dried slowly. Stalk yields of 20-30 tons/a are possible.

USES: The leaf stalk is the better known part of the plant in America, but the leaves and seeds are much used in Continental Cookery, where the leaves are added at the last moment to broth, soups, stews and stuffing. Leaves and stalks may be added to green salads. The French prepared from the seed, here used as a spice, the liqueur called Creme d'Celery, which does not differ greatly in flavor from kummel. The recipe may contain celery seed, plus or minus caraway, cumin or fennel, crushed and steeped in sugared vodka. Anise may be added but with care or it will be the dominant flavor.

FOLK MEDICINE: Regarded as aperient, carminative, diuretic, emmenagogue, nervine, sedative, stimulant, stomachic, and tonic, celery has been recommended for arthritis, bronchitis, catarrh, dropsy, frigidity, hernia, hypertension, hysteria, impotency, jaundice, kidney, lumbago, lungs, nervousness, melancholy, nerves, overweight, rheumatism, sciatica, spleen and uterus. A yellowish oil derived from the roots is said to repair sexual impotence brought about by illness. A rheumatism remedy recommended in the old herbals sounds pretty potent. It combines the "aphrodisiac" celery with the "aphrodisiac" damiana and the narcotic coca, source of cocaine. That should at least take one's mind off the rheumatism. Celery juice has been reported to lower blood pressure in hypertensive humans.

CORIANDER
(Coriandrum sativum L.)

CULTURE: Sow seed of this aromatic *annual* (ca 15 lbs/a) as soon as the ground can be worked in spring, marking the rows of the slow-germinating seeds with a fast germinator like basil or radish. Coriander is frost hardy and one of the first herbs to flower in spring. Seeds you don't collect this year, may germinate next year. Fairly easily grown in a window box, it is not so easily transplanted. Coriander leaves don't dry well, but can be frozen or steeped in salted oil. Yields of 1,000-3,000 lbs seed/a are possible.

USES: The renowned cilantro of Latin America, coriander leaves figure prominently in some Spanish dishes, chutney, salads, and soup. Ethiopians add the leaves to breads, sauces and teas. The seeds, considered aphrodisiac, become spicier as they dry. According to Mrs. Grieve, the seeds become narcotic if used too freely. Used in breads, cakes, pickles, sausages and tobacco mixtures, they are the secret ingredients in at least one type of whiskey and vermouth. The seeds and oil are widely used in bitters, gin, liqueurs and vermouths. Like ginseng, coriander is supposed by some not only to have aphrodisiac properties (of The Thousand and One Nights), but to impart longevity. Hence I would add some ginseng and gotu cola to John Farrell's Cinnamon and Coriander Cordial. The four herbs all used in the orient as aphrodisiac, mixed in about equal portions to flavor sugared vodka, plus or minus brandy, might counter the anaphrodisiac effects of the alcohol. Coriander, being chewed to alleviate bad breath, may mask the odor of the alcohol, offensive to some noses. Iranians smoke the fruits for toothache. Perhaps, because of the old belief that coriander imparted immortality, it is found among funeral offerings in Egyptian tombs. Russians extract linalol as a commercial starter for other chemurgics from the essential oil of coriander.

FOLK MEDICINE: Regarded as antispasmodic, aperient, aphrodisiac, aromatic, carminative and stomachic, coriander has been suggested for colic, erotomania, erysipelas, heart, hernia, indigestion, measles, nausea, rheumatism, scrofula, spleen, stomach, toothache and ulcers. Coriander is hypoglycemic in experimental animals.

CUMIN
(Cuminum cyminum L.)

CULTURE: More cold sensitive than cicely and coriander, cumin is a tender *annual* originally from the Mediterranean climate. It should be started indoors or in place after danger of frost is past. It takes four warm months to mature the seed.

Sown outdoors at rates of one to two seed per inch (or 25-35 lbs/a), plants are not thinned as the thick stand better holds the seed off the ground. When the fruits start browning, cut and dry indoors. Seed yields run 400-900 lbs/a (Duke, 1978; Rosengarten, 1973).

USES: Seeds are used to flavor breads, cheeses, cookies, fish, game, liqueurs, meats, sausages and vegetables. Ground seed are important ingredients in chili and curry powders. When pepper was priced out-of-sight, the Romans substituted cumin. Cumin finds its way into some of the recipes for Kummel, a European liqueur. It is the first seed mentioned in Time-Life's Seven Seeds Liqueur; ¼ cup of cumin, and ¼ cup each of angelica, anise, caraway, coriander, dill and fennel seed, crushed and added to 7 quarts eau de vie, with 6 cups sugar dissolved in 2 cups water.

FOLK MEDICINE: Ancients took the ground seed medicinally with bread, water, or wine, and it was considered the best of condiments. Smoking the seed results in a pallid complexion for which effect they were sometimes smoked in days of old. Leung (1980) mentions that the herb is used as an antispasmodic, aphrodisiac, carminative, diuretic, emmenagogue and stimulant.

DAMIANA
(Turnera aphrodisiaca Willd.)

CULTURE: So far I don't know how to raise this woody *perennial,* but I assume from its habit that it would require a dry tropical environment. I'd guess it could be raised like rosemary, bringing under glass when there's danger of frost.

USES: Damiana leaves are regarded by the counterculture as an "upper" to be taken for nervous or sexual debility. It has attracted quite a following as a reputed aphrodisiac. Sensing this, Mexican enterpreneurs make a commercial liqueur called Damiana. By boiling the leaves with sugar water, a tisane is generated, which can be concentrated to a syrup. Adding the syrup to vodka creates a homemade Damiana which can be improved by blending with other herbal syrups. The herb is also smoked, alone or in concert with other fumitory herbs, by the counterculture.

FOLK MEDICINE: Regarded as aphrodisiac, diuretic, purgative and stimulant tonic, damiana, said to act directly on the reproductive organs, has been used for cough, dysmenorrhea and nephritis. Heinerman (1979) mentions its use for bronchitis, emphysema, hot flashes and nervousness. Tyler (1982) suggests that damiana, as an aphrodisiac at least, is no more than an herbal hoax.

ELECAMPANE
(Inula helenium L.)

CULTURE: This huge *perennial* herb, to 6 feet tall, can be started from seed, faring well in ordinary garden soil in moist shaded circumstances. Root divisions, each with an eye, can be reset in the fall. Weeding and tillage both benefit the plant. Roots can be harvested after two years, drying some for your liqueur, while stratifying 2-inch fragments in sand, to start new plants for next spring. Yields of 3,000 lb/a dried root are possible.

USES: Roots have been used in aromatic bitters, confectionary, fold medicines, liqueurs, and vermouth. The root can be served macerated in water as a tea, or steeped in red wine. A cordial is made by soaking roots of elecampane with sugar and currants in white port. This cordial is recommended for colic. Elecampane is one of the richest sources of inulin, important in diabetic and other diets. Culpepper adds "The roots and herbes beaten and put into new ale or beer and daily drunk, cleareth, strengtheneth, and quickeneth the sight." The cordial of root steeped in wine is supposed to strengthen loose teeth, prevent their decay, and destroy worms, as well as allay pains of gout and sciatica. In Switzerland and France, elecampane is one of the ingredients of absinth. Also it enters the

French Vin d'Aulnee. Alantolactone inhibits germination and seedling growth, bacteria, fungi and worms. Further it lowers the blood pressure in experimental animals.

FOLK MEDICINE: Considered alterative, anthelmintic, antiseptic, cholagogue, diaphoretic, diuretic, expectorant, stimulant, and tonic, elecampane has been recommended for asthma, bronchitis, cancer, catarrh, coughs, consumption, diabetes, diarrhea, diptheria, dropsy, dyspepsia, gout, indigestion, itch, menstrual problems, neuralgia, phthisis, piles, respiratory inflammations, scabies, sciatica, spleen, stomach, urinary inflammations, and whooping cough. It has been recommended for skin diseases in horses, hence "horse-heal," and for scab in sheep, hence "scabwort." The infusion is reported to be sedative, at least in mice.

FENUGREEK
(Trigonella foenum-graecum L.)

CULTURE: Sow seed of this *annual* in the fall farther south, in the spring farther north, about 3 inches in the rows which may be 1½ ft. apart. Weeding is almost mandatory, as the fenugreek competes poorly, at least at weedy Herbal Vineyard. Seeds ripen 3-5 months after planting, and do not shatter (fall out of the pod) like so many other legumes. Planted at 15 to 25 pounds seed per acre, the plant yields 300-3000 pounds per acre (Duke, 1978; Rosengarten, 1973)).

USES: Dressed with oil and vinegar, the sprouts and seedlings are said to make a five-star salad. The seeds, containing coumarin, are used to flavour soups and salads. In Greece they are boiled and eaten with honey. In Lebanon, a milkshake-like hypotensive beverage is made by grinding the green seed after soaking. Also containing a yellow dye, the seeds are used cosmetically and medicinally, but more importantly as a constituent of Indian curries. Seeds are used as a coffee substitute (Watt and Breyer-Brandwijk, 1962). Leaves and young pods may be cooked as vegetable, but they are bitter. They are said to be extremely rich in vitamin C. Fenugreek butters and teas are reported. With its coumarin flavor, fenugreek should make a good herbal vinegar. The seeds are used in making imitation maple syrup. In veterinary medicine, it is mixed with bad tasting hay or medicine to induce the animals to eat. The dried plant is sometimes mixed with stored grain as an insect repellent. With its bitter taste, somewhat like lovage and celery, it might be useful in vegetarian bouillon. For northerners, attempting to make liqueurs which call for tonka or vanilla, fenugreek seed may be used as a substitute. According to Tyler, this is one of the major ingredients, after alcohol, in Lydia Pinkham's compound. Recently fenugreek has proved to be a useful source of the drug diosgenin, used in the synthesis of hormones. As with other chemurgic crops, a large percentage of the crop must be thrown into the pot to extract a small percentage (1-2 percent on a dry-weight basis) of pharmaceutical (diosgenin). While the "pot is boiling," proteins, fixed oils, oleoresins, e.g., coumarin, mucilages and/or gums might be extracted. Organic residues might be used for biomass fuels or manures, inorganic residues for "inorganic" chemical fertilizers. The husk of the seed might be removed for its mucilage, with the remainder partitioned into oil, sapogenin, and protein-rich fractions (Duke, 1981).

FOLK MEDICINE: Fenugreek shares with *Cannabis* and several other seeds, urchins and jellyfish, the alkaloid trigonelline. Coincidentally both fenugreek and *Cannabis* are said to enlarge the bust. Oil of fenugreek seed contains a lactation-stimulating factor but no estrogenic, testosterone-like or progesterone-like activity could be demonstrated. On a dry basis the seed contains 0.1 percent diosgenin, 0.01 percent gitogenin and a trace of tigogenin. In folk medicine seeds are considered aperient, aphrodisiac, carminative, demulcent, diuretic, ecbolic, emollient, galactagogue, insecticidal, stimulant, and tonic. In East Africa, macerated seed are made into a porridge with *Carum copticum* and *Cuminum cyminum* to treat diarrhea. In Egypt macerated seed infusions are as highly regarded as quinine for fever, and the seeds also find their way into folk remedies for anemia, gout, rickets, and scrofula. Other medicinal uses, e.g. in Lebanon, where the plant is considered a panacea, are detailed in Duke's Medicinal Plants of the Bible.

GINSENG
(Panax quinquefolius L.)

CULTURE: *Perennial* from seeds or small plants grown from seed. Seeds ripen in the fall but usually do not germinate until the following fall, or more commonly in Maryland, two springs later. Hence seeds must be stratified until used. Stratification consists of storing the seeds in a cool, moist place, using forest soil, sand, loam or sawdust as a storage medium. Seeds sown into the permanent beds may be set about 8 inches apart each way. If sown in a seedbed for subsequent transplanting, 2-6 inches is sufficient. Seedlings are transplanted after 2 or 3 years and set at 8 inch intervals. Seedlings purchased from a supplier may be put directly into permanent beds. Good drainage and rich soil are desirable. Beds should be about 4 feet wide to provide easy access and drainage in the walkways. Lathe shading may be provided to allow about 75 percent shading. Free circulation of air is desirable. Fertilizing with good forest soil or loam is suggested; most growers advise against use of chemical fertilizers or barnyard manure. Mulching is needed in most plantations in winter or during dry periods. Yields 3-7 years after planting may range from ½-1 ton dried root per acre from seed.

USES: Ginseng is one of our biggest herbal exports, exceeding 35 million dollars worth per year recently. Ginseng is one of the most popular items, with aloe and jojoba, in natural cosmetics of all sorts. It is one of the main components in a tea I call root booster, with ginger, sarsparilla, and sassafras; with the addition of a little gin or vodka, this makes a very acceptable liqueur, especially if tinted red with bergamot petals. One of my favorite liqueurs is gin, ginseng, and ginger, tinted green, and I call it the Green Giant. While I have not seen much American ginseng in liqueur, I saw dozens of kinds of wines, tonics, and liqueurs with ginseng roots steeping therein in both Korea and China. Recently the FDA has suggested that it is legitimate to add water to ginseng (water then being an approved food additive) but not the ginseng to water (water is a food but ginseng is not approved as a food additive or a drug). These days I'm not as sure as the FDA that city water or well water is salubrious.

97

FOLK MEDICINE: Root used as a panacea (cure-all) by wealthy Chinese and other orientals, who consider it carminative, diuretic, stimulant, and tonic. Chinese sick chew the root to recover health. Healthy Chinese chew it to increase their vitality. It is said to remove both mental and bodily fatigue, to cure lung disorders, dissolve tumors, and prolong life. Reportedly it reduces blood sugar concentration and acts favorably on metabolism, the central nervous system and on the endrocrine secretions. Employed in the Orient in the treatment of anemia, diabetes, insomnia, neurasthenia, gastritis, and especially for sexual impotence. The American scientific community recognizes only its demulcent properties, although Applachian folk uses include tonic and aphrodisiac. An infusion of the leaves is said to make a palatable tea. Orientals attribute different medicinal virtues to the endangered oriental gingeng and our occidental ginseng.

HOREHOUND
(Marrubium vulgare L.)

CULTURE: Said to be difficult to start from cuttings or root divisions, this *perennial* is one of the easiest herbs to start from seed. Young seedlings should be thinned to 12 inches apart. They soon become difficult to transplant. Tolerating hot, dry, poor soils, horehound requires plenty of sun. Yields of 2,000-3,000 lbs leaf per acre are possible.

USES: The dried herb is used in making horehound candy and horehound ale as well as flavoring cakes, salads, sauces, and stews. A vodka liqueur made by steeping horehound should help in bronchitis, colds, and coughs. Horehound ale is a salubrious aperient. The extracts are used in making bitters and liqueurs.

FOLK MEDICINE: Regarded as alexeteric, antispasmodic, aperient, diaphoretic, diuretic, expectorant, stimulant, stomachic, tonic, and vermifuge, horehound has been recommended for asthma. bronchitis, cough, consumption, heart palpitations, hoarseness, itch, jaundice, nerves, paratyphoid, snakebite, and typhoid. I recommend a hot concoction of horehound with vinegar, hot pepper and horseradish, to take at bedtime for sore throat. Marrubiin, a compound in hore-hound, does in fact have expectorant properties, breaking up the throat congestion. After chemical modification (when its lactone ring is opened) it exhibits strong choleretic activity. Homeopaths will be pleased to read in Leung (1980) that in large doses it causes arrhythmias of the heart while in smaller doses it tends to correct them.

HYSSOP
(Hyssopus officinalis L.)

CULTURE: Easily grown in a light soil with sunny exposure, this hardy *perennial* can be raised either from seed or cuttings, made in early spring or autumn.

USES: Culinarily used for soups, teas, and liqueurs. A hyssop tea, mint-flavored as it is, is recommended for asthma, colds, and rheumatism. It may be mixed with horehound. Hyssop baths have also been prescribed for rheumatism. Hyssop is much employed in the making of liqueurs, being an important element in chartreuse, and in some absinths. It is also used in bitters.

FOLK MEDICINE: Regarded as astringent, carminative, diaphoretic, emmenagogue, expectorant, pectoral, stimulant, stomachic, tonic, hyssop has been recommended for asthma, bronchitis, bruises, burns, catarrh, cold, cough, dropsy, in-digestion, infections, inflammations, jaundice, rheumatism, scrofula, sore throat, tumors and worms. The extract is said to be antiviral against herpes simplex.

LEMON BALM
(Melissa officinalis L.)

CULTURE: Once started, and easily started from cuttings, root divisions or seed, this *perennial* tolerates many insults, shade, sandy soils, waterlogging. Situations that promote lush growth produce a foliage low in aroma. The rougher you treat it, the better it treats you, aromatically speaking. Yields of 1,000-2,000 lb/a leaf are easily attainable.

USES: The lemon-scented leaves find their way into all sorts of culinary and medicinal concoctions. I use it as a lemon substitute in both hot and iced teas. Balm tea, itself, sweetened with honey, was supposed to impart long life. Leaves are also added, sparingly to salads. An old fashioned balm claret recipe calls for a quart of claret, a pint of soda water, a shot glass of cognac, sweetened, plus a bunch of balm and borage, and half-a-cucumber in sections. Zanelli (1972) uses 2 quarts of balm leaves, 3½ lb sugar, ¾ oz. citric acid, a pinch of tannin, 1 gal. water, a dash of Tokay yeast and 5 days aerobic fermentation for his Balm Wine. Farrel (1974) makes his Melissa Liqueur using sweetened brandy. I prefer to boil down a fresh plant with sugar to form a thick syrup to which I add four or five parts gin or vodka, both considerably cheaper than brandy where I come from. Farrel recommends his Melissa Liqueur with after-dinner coffee.

An essence of balm in Canary Wine "every morning will renew youth, strengthen the brain, relieve languishing nature, and prevent baldness." Another old remedy, for nervous headache and neuralgia embraced balm, angelica, nutmeg, and lemon-peel. This was called Carmelite Water, and I term a liqueur made by blending these with vodka Carmelite Liqueur. Carmelite Water was consumed daily by Emperor Charles V. Five days aerobic fermentation of 1 gallon balm leaves, 7 lbs sugar, in 2 gals water with 1½ oz citric acid and a dash of tannin is said to produce a respectable wine. The leaves can substitute anywhere lemon peel is recommended, e.g., as the twist of lemon in a martini. Balm is included in some recipes for Benedictine and Chartreuse.

Foster (1973) suggests rubbing fresh balm on wooden furniture, not only to impart aroma and sheen, but to discourage felines from clawing the furniture. Crushed leaves are poulticed on sores, tumors and insect bites. Balm flowers are a bee favorite. If hives be rubbed with the leaves, it's said the bees will stay together.

FOLK MEDICINE: Regarded as antispasmodic, calmative, carminative, diaphoretic, emmenagogue, and stomachic, balm has been recommended for asthma, bronchitis, catarrh, colds, colic, cramps, dizziness, dysmenorrhea, dyspepsia, fever, flu, gastroenteritis, gout, headache, heart, herpes, hysteria, indigestion, insomnia, leucorrhea, melancholy, menopause, migrain, mumps, nausea, nerves, neuralgia, nightsweat, sores, stings, stomach, toothache, tumors, and vertigo. According to Leung, hot-water extracts (equal lemonbalm tea) are antiviral against herpes, mumps, newcastle's disease and other viruses. The oil is bactericidal, and, in experimental animals, antihistamic and antispasmodic.

LOVAGE
(*Levisticum officinale* W.D.J. Koch)

CULTURE: Best started from fall-planted seed for transplant in the spring, this is a tall *perennial*. In northern latitudes, full sun is recommended, in southern latitudes, partial shade. In either case, rich well-drained soil is desirable. In the fall roots may be subdivided, insuring that each portion has an eye. Clumps should be divided every two or three years. Yields of 1,500 lb/a dry root are possible.

USES: Finding its place instead of celery in many recipes, the leaves, seeds, and roots provide the celery flavor. The stems are candied like angelica, the seeds spread on bread and biscuits like caraway. Lovage was once worn as a love symbol and was put in love potions as a guarantee of everlasting devotion. Oil from the plant is used for flavoring fumitories like tobacco. There's a cordial called lovage, containing lovage, tansy, and yarrow. Lovage was popular in Britain before World War I. Two tablespoons of macerated leaves in a quart of brandy with a cup of sugar and a dash of dried tansy (poisonous in quantity) and yarrow was the recipe for one lovage liqueur. Dried leaves are sutiable for making teas and liqueurs. Crushed leaves are applied to beestings.

FOLK MEDICINE: Regarded as aromatic, carminative, diuretic, emmenagogue, expectorant, stimulant, stomachic, lovage has been recommended for ague, boils, catarrh, colic, dysmenorrhea, flatulence, gravel, jaundice, pleurisy, quinsy, skin, sore throat, and urinary disorders. Since lovage promotes the onset of menstruation it should not be used by pregnant women.

ORRIS
(Iris florentina L.)

CULTURE: Rhizomes of this *perennial* are divided in late summer or early fall to be set out in enriched well-drained sunny situations. New clumps must be weeded carefully.

USES: Primarily used in perfumery, cosmetics and face powder. Noting "It is not an edible herb," Foster (1973) goes on to talk about sweetflag and its edible parts so that careless readers may get confused. The fresh root of orris is in fact a powerful cathartic. Extremely acrid when fresh, the roots when carefully dried are almost sweet, with the aroma of violets. In the past, the "poisonous" orris was used to flavour brandies. In Russia it was mixed with ginger and honey to make a beverage which of course could be fermented or spiked to make a cordial of liqueur. Due to its potency no more than one tablespoon should be steeped for a week or so in brandy or vodka.

FOLK MEDICINE: Regarded as cathartic, diuretic, emetic, sternutatory, and stomachic, orris has been recommended for bronchitis, colic, coughs, diarrhea, dropsy, halitosis, liver congestion, and sore throat.

REQUIENI MINT
(Mentha requienii Benth.*)*

Delicate mat-forming *perennial,* this exceedingly aromatic mint is a wonderful plant to encourage around your stepping stones. Like most mints, it is easily propagated by cuttings.

This is the essential element of Creme de Menthe. Further it makes a nice tea, blended with lemon balm. Peppermint, the mint of Mint Juleps, and Spearmint, **perennial** ''weeds'' once established, are also useful in herbal liqueurs. Surprisingly, I have no data in our computer for folk medicine uses of this noteworthy mint.

ROMAN CHAMOMILE
(Anthemis nobilis L.*)*

CULTURE: Flourishing with little attention in dry, sandy, sunny situations, this *perennial* may be propagated by seed, sown shallow in the spring, or from subdivisions (as many as 50) the following spring. Lime is to be avoided. The more desirable double-flowered forms (the flowers are harvested) fare better in a moist black loam. Tends to volunteer and may become weedy. British have taken advantage of this and let it go to turf. The bitter flowers are picked as they appear, and placed in a paper bag to dry. Flower yields of 500 lb/a are possible.

USES: One of the most popular sedative teas in Latin America, manzanilla serves as an after-dinner beverage, especially in countries less addicted to pills.

Wine can be made with the flowers, sugar, citric acid, and sauterne yeast, following 5 days aerobic fermentation. A stomachic liqueur is made by boiling the flowers in sugar water with ginger and perhaps other spices, and adding to gin or vodka. The whole plant is used in making bitter herbal beers, and the essential oils or extracts in Benedictine liqueurs, bitters and vermouths.

The tisane is used in lotions for the body, to remove weariness of aching joints, and for the hair, to brighten and/or soften blond hair. Old herbals say chamomile is 120 times as antiseptic as seawater. It is widely used in the old world, as a collyrium, cleansing astringent, sleep inducer, and tranquilizer.

FOLK MEDICINE: Regarded as anodyne, antispasmodic, aperient, bitter, diaphoretic, diuretic, sedative, stimulant, stomachic, and tonic, manzanilla has been recommended for bladder, colic, cramps, delirium, dropsy, dyspepsia, eczema, erotomania, eyes, flatulence, gastroenteritis, gout, headache, hysteria, indigestion, inflammation, insomnia, jaundice, kidney, leucorrhea, lumbago, malaria, nausea, neuralgia, nightmares, pain, piles, sciatica, scrofula, spasm, stomach, toothache, tumors, typhus, worms, wounds.

ROSEMARY
(Rosmarinus officinalis L.)

CULTURE: Although this woody *perennial* can be started from seed, it readily roots as cuttings. It does very well in hanging pots by exposed kitchen windows. Heavy frosts kill the plant, and it is difficult to overwinter outdoors in Maryland.

USES: A **perennial** to 2 m tall, this evergreen blue-flowered shrub is native to the Mediterranean. As a camphoraceous spice, it is used with chicken, duck, fish, lamb, pork, rabbit, shellfish, soups, stews, and veal. Among vegetables, is it used with eggplant or potatoes, or zucchini but sparingly. Harrup (1977) offers a recipe for rosemary bread. Some cultivars are adaptable to bonsai and hanging gardens. Foliage yields of 1,000 lb/a seem feasible.

Rosemary finds her way also into herbal baths, herb biscuits, bouquets, potpourris, weddings, and even funerals as a symbol of fidelity. As a cordial, it is reputed to stimulate the blood, eyes, and memory. The Rodale Herb Book (Hylton, 1974) suggest rosemary, germander and lemon verbena for the basis of one herbal liqueur. Zanelli (1972) does not recommend rosemary as a basic wine ingredient, although it is sometimes used in combination. (Do not use more than 1 oz per gallon of water). Rosemary, lavender, myrtle, steeped in vodka, makes a satisfactory liqueur, called Hungary Water, said to cure paralysis. It was once massaged onto gout of the hands and legs. As a massage ointment, rosemary spirits are said to prevent premature baldness. A cold mixture of dried rosemary mixed with borax is said to be a good dandruff shampoo. Boiled in white wine, the leaves constitute a facial lotion, believed to make fair the complexion. I find it interesting that this antiwrinkle plant has proven out to contain an antioxidant known as rosmarol. The leaves are said to enhance the flavors of wine and to prevent spoilage. Rosemary tea is too camphoracous for my taste, but it makes a nice additive to teas and liqueurs.

Mrs. Leyel says "A tisane of Rosemary will cure a nervous headache and has a beneficial effect on the brain. Its constant use will greatly improve a bad memory." The tea is also good for colds, colic, and nervous tension. The Rodale Herb Book suggests perfuming the house by placing a pan of water with crushed rosemary and juniper on the radiator. The French are said to burn the mixture of rosemary and juniper berries. "Herbal" Ed Smith makes his Amazonian bug repellent out of 1 part each of rosemary, rue, wormwood and basil (Conrow and Hecksel, 1983). All these herbs are reported to have insect repellent qualities and probably allergenic properties as well, especially the rue. Rubbed together with coltsfoot leaves, rosemary is smoked to treat asthma, lungs and throat conditions. Farrel suggests adding bruised rosemary to Scotch to simulate Drambuie. The oil is used in cosmetics, deodorants, liniments, lotions, soaps and tonics, as well as liqueurs and medicinals.

FOLK MEDICINE: Medicinally rosemary is variously considered antispasmodic, aperient, carminative, dissolvent, diuretic, ecbolic, emmenagogue, febrifuge, nervine, stimulant, stomachic.

SWEET CICELY
(*Myrrhis odorata* L. Scop.)

CULTURE: With the odor of lovage, the flavour of anise, this tall hardy *perennial* should be sown when the seed ripen in late summer, as the seed take about nine months to germinate. It's good on a northern exposure, not receiving sun all day. Rich moist soil is recommended. Seed yields of 1,000-2,000 lbs/a seem feasible.

USES: Roots were once boiled for use as a vegetable or in salads. The leaves may be used as an aromatic garnish like parsley, and they serve well with salads, seafoods, soups and stews. The leaves might serve as a dietary sugar substitute. The dry seeds are used in cakes and cookies, and can be used like coriander, anise, or fennel in liqueurs. A tisane made of the leaves aids digestion. Green seeds, a pleasant nibble in themselves, are added to salads like capers, and the seeds are used as a substitute for aniseed in many spicy dishes and liqueurs. The liqueur I call Nicely Cicely might be popular with old and young alike. One old herbal notes that cicely seeds, boiled and served with oil and vinegar and "very good for old people that are dull and without courage; it rejoiceth and comforteth the heart and increaseth their lust and courage." A decoction of the roots, on the other hand, is said to be a good tonic for girls 15 to 18 years old (Grieve, 1931). Since the leaves taste of sugar, they can be used to cut down on the sugar in recipes, including those for liqueurs. Nicely Cicely is made simply by steeping macerated leaves and/or roots of cicely in lightly sugared vodka or gin, with or without the addition of other alleged aphrodiasiac or fertility herbs such as absinth, basil, caraway, celery, damiana, fenugreek, ginseng, lovage, nutmeg, savory, and yohimbe. The cicely can provide both the sugar and anise in several possible recipes. The oil from the seeds was once used to polish English oak floors.

FOLK MEDICINE: Regarded as aperient, aphrodisiac, aromatic, carminative, digestive, expectorant, stomachic, and tonic, cicely has been recommended for consumption, cough, flatulence, indigestion, sores, stomach.

Table 8.
GRAS Liqueur Ingredients (after Leung, 1970)

Absinthium	Chirata[3]	Marjoram
Alfalfa	Cinchona[5]	Mint
Allspice	Cinnamon	Myrrh
Almond	Citronella	Nutmeg[2]
Aloe	Clary	Olibanum
Althaea	Clove	Onion
Ambrette	Coca[6]	Orange
Angelica	Cocoa	Oregano
Angostura	Coffee	Origanum, Spanish
Anise	Coriander	Papain
Artichoke	Cornsilk	Parsley
Ash, Prickly	Costus	Passionflower
Asparagus	Cubebs	Patchouly
Balm	Cumin	Pepper[2]
Balsam Copaiba	Damiana	Peppermint
Balsam Peru	Dandelion	Pine Bark[3]
Balsam Tolu	Dill	Pine Needle
Basil[2]	Dill, Indian	Pipsissewa
Bay	Elder Flowers	Quassia
Bay, West Indian	Elder Leaves[4]	Quebracho
Beet	Elecampane[3]	Quillaia
Benzoin	Elemi	Rhubarb[3]
Bergamot	Eucalyptus	Rose Oil
Birch Oil[1]	Fennel	Roselle
Blackberry	Fenugreek	Rosemary
Black Haw	Galbanum	Rue
Bois de Rose	Genet	Saffron
Boldo	Gentian	Sage
Boronia	Geranium	Sandalwood
Buchu	Ginger	Sarsaparilla
Cajeput	Grapefruit	Sassafras[7]
Cananga	Guarana	Savory
Caraway	Guaiac	Spearmint
Cardamon	Hops	Storax
Carob	Horehound	Tamarind
Carrot	Hyssop	Tarragon
Cascarilla	Immortelle	Tea
Cassia	Jasmine	Thyme
Cassie	Juniper	Tumeric
Catechu	Kola	Turpentine (Rosin)[3]
Cedar Leaf[1]	Labdanum	Valerian
Celery	Lavender	Vanilla
Centaury[3]	Lemon	Woodruff[3]
Chamomile	Lemongrass	Yarrow[2]
Cherry Bark[4]	Licorice	Yerba Santa[3]
Cherry Laurel[4]	Lime	Ylang Ylang
Chestnut Leaf	Lovage	Yucca
Chicory	Mace[2]	

[1] Only if thujone free. [2] May contain safrole. [3] for alcoholic bev. only.
[4] provided HCN does not exceed 25 ppm. [5] Provided total cinchona alkaloids do not exceed 83 ppm in final beverage. [6] If decocainized. [7] If safrole free.

LIST OF LIQUEURS
(and one or more herbs contained therein)

Abricotine: apricot (France)

Absinthe: angelica, anise, coriander, elecampane, fennel, wormwood.

Advocaat: brandy, egg yolk, sugar (Holland)

Aguardiente: anise (Latin America)

Aiguebelle: more than 50 herbs (France)

Akvavit: caraway

Alkermes: brandy, cinnamon, cloves, coriander, jasmine, nutmeg, orange flowers, orris, rose, vanilla (Italy)

Allasch: almond, anise, caraway (Latvia)

Allspice: allspice

Amer Picon: orange, quinine, spices (France)

Anesone: anise

Angelica: angelica and other herbs from the Pyrenees, e.g., anise, balm, cardamon, cinnamon, clove, hyssop, mace, nutmeg, marigold

Angostura: cinnamon, clove, cusparia, lemon, mace, nutmeg, orange, rum

Anis del Mono: anise (Spain)

L'Anisette: fennel (France)

Anisette: almond, angelica, anise, coriander, fennel, star anise.

Apricot Liqueur: apricot

Apry: apricot (France)

Aquavit: caraway (Scandinavia)

Arak: licorice (Turkey)

Armagnac: brandy, black oak casks (France)

Arrack: rice a/o palm wine (Scandinavia)

Aurum: oranges and herbs from Abruzzi mountains (Italy)

Baerenfang: honey, linden, mullein

Balsam: banana (West Indies)

Barack: apricot (Hungary)

Benedictine: more than 50 herbs, angelica, allspice, balm, hyssop, peppermint, sweetflag, thyme (France)

B & B: drier than Benedictine

Bishop: allspice, cinnamon, clove, ginger, lemon, mace, nutmeg

Bitters: angostura, cusparia

Blackberry: blackberry
Black-currant: black currant
Brazilia: coffee (Brazil)
Bronte: brandy, herbs, honey (England)
Buchu: buchu, brandy (South Africa)
Byrrh: quinine
Cacao mit Nuao: chocolate, filberts (Germany)
Calisay: quinine (Spain)
Calvados: apple (Norway)
Carlsbery: herbs, mineral water (Germany)
Carmelite water: angelica, balm, lemon-peel, nutmeg
Cayo verde: lime (U.S.)
Celery Cordial: caraway, celery, fennel
Centerbe: more than 100 herbs (Italy)
Cerasella: cherry liqueur (Italy)
Channelle: cinnamon and other spices
Chartreuse, green: more than 230 botanicals, among them
   allspice, angelica, balm, basil, bergamot, cicely, coriander,
   hyssop, orangemint, peppermint, rosemary, sweetflag,
   tansy (France)
Chartreuse, white: angelica
Cherry Blossom: "aroma" of cherryblossoms (Japan)
Cherry liqueur: cherry
Cherry heering, small: Danish black cherries (Denmark)
Cherry marnier: cherries (France)
Cherry suisse: cherry and chocolate (Switzerland)
Claarava: herbs and honey (Scotland)
Claristine: herb-flavored liqueur (France)
Cocuy: sisal root brandy (Venezuela)
Coffee liqueurs: caramel, cardamon, cinnamon, clove, coffee,
   orange, vanilla.
Coinguarde: quince (France)
Cointreau: green oranges and brandy (Curacao)
Cordial Medoc: claret, cognac, curacao orange (France)
Creme de Ananas: brandy, pineapple, vanilla
Creme d'Angelique: angelica, cinnamon, clove
Creme de Banana: banana
Creme de Cacao: cacao, cloves, mace, vanilla
Creme de Cafe: coffee
Creme de Cassis: black currant (France), cinnamon, clove
Creme de Celeri: anise, caraway, celery, fennel

Creme de Cerise: cherry
Creme de Cumin: cumin, sugar
Creme de Fraises: strawberry
Creme de Framboise: raspberry
Creme de Genieve: juniper
Creme de Macis: mace in vodka
Creme de Menthe: cinnamon, corsican mint, cumin (Fox, 1933), ginger, orris, pennyroyal, peppermint, sage
Creme de Moka: coffee beans, brandy
Creme de Muscade: nutmeg
Creme de Noyau: apricot kernels, peach
Creme de Recco: brandy, sugar, tea leaves (Italy)
Creme de Roses: rose petal ± rosehip
Creme de The: brandy, sugar, tea leaves (France)
Creme de Vanille: alcohol and vanilla
Creme de Violets: violet petals
Creme de Yvette: violet (U.S.)
Cristal Floquet: orange
Curacao: cinnamon, clove, curacao orange peel (Holland)
Curanta y tres: 43 herbs (Spain)
Damiana: damiana
Danzig Cordial: distilled buckwheat beer
Danzigwasser: orange or lemon peel, anise
Delecta: herbs
Dewmiel: scotch
Drambuie: scotch, heather, honey and herbs, anise, angelica
Eau de Carnes: (France) coriander
Eau de melisse: balm
Eau de Noyaux: almond
Eau de Vie: brandy of whatever fruit
Eau de Vie d'Hendaye: anise
Elixir d'Anvers: bittersweet liqueur (Belgium)
Elixir d'Bacardi: rum base (Cuba)
Elixir d'China: staranise (Italy)
Eltaler: herb-flavored liqueur (Germany)
Enzian: gentian et al (Germany, Austria)
Escarchado: anise, sugar (Portugal)
Expresso: coffee
Falernum: almond, lime
Fenouilette: fennel
Fiore d'Alpe: edelweiss ± other alpine herbs

Fleur de Mocha: coffee
Flora Alpina: edelweiss, et al.
Forbidden fruit: brandy, shaddock, valencia orange
Fraise: strawberry
Framboise: raspberry
Galliano: herbs, spices (Italy); angelica
Gallweys Irish Coffee: coffee, Irish, herbs, honey
Geneva Gin: juniper (French)
Gilka: caraway et al (Germany)
Gin: angelica, coriander, juniper
Ginger Cordial: cardamon, cinnamon, clove, ginger
Ginger-flavored brandy: brandy, ginger, pepper
Glayva: herbs and spices (Scotland)
Glen Mist: scotch, honey, etc.
Goldwassar: caraway, lemon or orange peel
Gonry Doubnyak: acorns, angelica, cloves, ginger, galingale
  (Russia)
Grand Marnier: cognac, orange
Grapefruit Liqueur: cinnamon, coriander, orange
Grappa: grapebrandy, rue
Grenadine: pomegranate
Guignolet: cherry (French)
Herbsaint: *no* wormwood (New Orleans)
Highland Bitters: chamomile, cinnamon, clove, coriander,
  gentian, orange peel
Hippocras: cinnamon, ginger, melegueta pepper
l'Huile de Venus: caraway (France)
Irish Mist: herbs, heather honey, Irish
Izarra: brandy and herbs of the Pyrenees
Jagermeister: a red liqueur (Germany)
Juniper Cordial: coriander, juniper
Kahlua: coffee (Mexico)
Kalmus: anise
Karpi: cranberry et al (France)
Kir: black currant, dry white wine (France)
Kirsch: cherry brandy
Kirschwasser: distilled fermented sweet cherries
Kola: citrus, cola, tonka a/o vanilla
Kona: coffee
Kummel: caraway, cumin, fennel, orange, orris
La Senacole: herb-flavored liqueur (France)

La Tintaine: anise, fennel (France)
Lindesfarne: honey, whiskey (English)
Liqueur des Moines: cognac, herbs (France)
Liqueur de Or: anise, caraway, citrus (France)
Liqueur de Sapin: herbs, spices
Lochan Ora: Chivas Regal + X
Lovage: lovage, tansy, yarrow
Luana: coffee
Luxardo: marasca cherry (Italy)
Maitrunk: woodruff
Mandarine: brandy, tangerine
Manzanilla Sherry: chamomile
Maraschino: marasca cherry
Marmot: cacao
Marnique: quince (Australia)
Mastic: anise, mastic resin (Greece)
Masticha: anise (Greece)
Mastiha: grape brandy, anise, mastic
May Wine Punch: benedictine, brandy, wine, woodruff (U.S.)
Mazarine: herbal liqueur (Argentina)
Medoc: brandy, claret, curacao orange
Melisse: balm
Metheglin: cinnamon, cloves, elder, ginger, marjoram,
    pepper, rosemary (Welch)
Mille Fior d'Alpi: alpine herbs
Mint Julep: spearmint
Mirabelle: plum
Mobana: bananas (Bahamas)
Monastique: brandy, herbs
Monte Aguila: allspice (Jamaica)
Noyau: apricot pits, peach pits
O Cha: green tea (Japan)
Ojen: anise (Spain)
Okhotnichya (Hunter's Brandy): aniseed, citrus peel, clove,
    coffee, galingale, ginger, juniper, peppers (Russia)
Okolehao, dark: coconut, rice, ti
Okolehao, white: cane, coconut, rice
Orange Curacao: cinnamon, clove, orange
Orange Liqueur: cinnamon, coriander
Orgeat: almond, orange blossoms
Orris: anise

Ouzo: anise (Greece)
Parfait Amour: violet petals, lemon peel, vanilla (France)
Passionfruit: passionfruit (Australia)
Pastis: anise (France)
Peach: peach (U.S.)
Peppermint Schnapps: peppermint
Perkom Brandy: St. John's wort (Norway)
Pernod: anise
Perry: pear
Pertsovka (pepper vodka): capsicum, cayenne, cubeb (Russia)
Pferfenmunze: anise (Germany)
Pimento Dram: allspice, rum (Jamaica)
Pisco: muscat grapes
Poire-Williams: pear
Prunelle: plums, prune (France)
Quetsch: plum
Quince-flavored Brandy: brandy, quince, sugar
Raki: anise oil, licorice, mastic (Turkey)
Raspail: angelica, calamus, myrrh (France)
Reishu: melon (Japan)
Rock & Rye: lemon, orange, rockcandy, rye whiskey
Roscan Aromatique: anise
Rosemary: balm, coriander, sage, thyme
Rosolio: rose petals
Rosolio de Torino: anise (Italy)
Rumfustian: cinnamon, ginger, nutmeg
Rumona: cinnamon, rum, tonka, vanilla (Jamaica)
Sack: mead, fennel
St. Hallvard: herbal liqueur (Norway)
Sambuco: licorice et al (Italy)
Sassafras cordial: balm, brandy, citron, juniper, lemon,
   marjoram, pistachio, rosemary, sarsaparilla, sassafras
Schnapps: caraway, peppermint
Seve: herbal liqueur (France)
Seven Seeds' Liqueur: angelica, anise, caraway, coriander,
   cumin, dill, fennel
Singaree: allspice, cinnamon, clove, nutmeg
Slivovitz: plum (Israel)
Sloe Gin: blackthorn (Britain)
Southern Comfort: bourbon, peaches (U.S.)
Stonedorfer: herbal liqueur (Germany)

Start: licorice
Strega: several herbs and spices (Italy); angelica
Suze: gentian bitter (France)
Swedish Punsch: palm wine, rice, lemon, spices, tea
(Scandinavia)
Tamara: date (Israel)
Tapio: herbs, juniper (Finland)
Thitarine: fig, licorice
Tia Maria: rum, coffee (Jamaica)
Tiddy: honey-sweetened liqueur (Canada)
Trappistine: herbal liqueur (France)
Tres Castillos: anise, rock candy (Peurto Rico)
Triple Sec: orange peel
Unicum: orange (Hungary)
Usquebaugh: cardamons, cloves, nutmeg, saffron (Irish),
anise, (English)
Van der Hum: herbs, tangerine (S. Africa)
Vandermint: cacao, peppermint (Holland)
Vanilla Bean Cordial: vanilla, vodka (France)
Vermouth: angelica, coriander, elecampane, russian
wormwood, sweetflag
Verveine du Verlay: herbal liqueur (France)
Vieille Cure: more than 50 herbs (France)
Waldmeister: brandy, wine, woodruff
Wassail (Yuletide): cinnamon, ginger, lemonpeel, nutmeg
Wishiowka: cherry (Russia)
Wishniak: cherry (Poland)
Zubrovka: buffalo grass

CHAPTER 5

'Quack' Salad[1]

Retiring at age 65, my father looked forward to a full life of golf, something his insurance career would not fully permit. Cancer cut him down less than a year later, just as it did at least two of his brothers, both shortly after age 65. He didn't smoke or drink, and he advised against such habits. Childishly I told him they'd have a cure for cancer before my time came. That was when I was around 20. He died when I was nearly 30. Still no cure or prophylaxis for all cancers was in sight. By the time I was 40 I feared my prediction had soured. There was no cure for most cancers. So I kicked the three-pack-a-day smoking habit. But the tars of 25 years of intensive smoking coated my lungs.

My father and his brothers died of cancer of the lower bowel. They had graduated from the fibrous farm diet they

[1] With permission of Mark Bricklin, The Practical Encyclopedia of Natural Healing. 1976, Emaus, Pa.

lived on as kids in the Deep South to the refined bread, meat, and potato diet of modern America. It seems generally accepted now that lack of fiber in the American diet correlates with cancer of the lower bowel. That's why you'll often find me eating unhusked cereals and seeds. I need the fiber that the food processing industry goes to such pains to remove (with considerable energy expenditure).

Repeating now that science has not yet come up with a cure for all cancer, I have resorted to folk medicinal lore which has given science such esteemed and powerful drugs as atropine, cocaine, henbane, ipecac, morphine and reserpine. Now I list 25 "salad" plants reported by Dr. Jonathan Hartwell (1967-71) of the National Cancer Institute as folk remedies for cancer.

atriplex	crimson clover	peanut
beet	cucumber	safflower
black walnut	cumin	sage
borage	flax	stinging nettle
calendula	garlic	tamarind
celery	hot pepper	tansy
chicory	licorice	tea
chive	onion	tomato
chufa		

Over half of these species possess some compound that has been useful in some treatment of some type of cancer. You'll find many others in the next chapter, Spices of Life.

Now I'll be the last one to say that tossing all these ingredients will cure or even alleviate cancer. I'll be the first to say, however, that you can make a decent salad combining these ingredients, using large portions of the standard salad ingredients and smaller portions of the strangers. I have sampled at least one leaf each of these plants and survived. I imagine I am helped more than hurt by having eaten these items. All plants manufacture important vitamins, and probably all manufacture some ingredients that are toxic if consumed in large quantities. Several thousand have been used in folk remedies for cancer, and many of these have scientific merit in the treatment of one cancer type or another.

I can't guarantee that the "quack" salad will cure or even prevent cancer. I imagine most people would be healthier if they ate such a salad once a day. I will surely nibble on these on my daily visit to my "quack" salad greenhouse. When time permits, I'll mix up the salad, using lemon juice, hot pepper, and garlic as a salad dressing. This is a dressing used by several Indian tribes of Latin America, among them the Cuna Indians of Panama. All three of the dressing ingredients show up in folk remedies for cancer.

If even 10 percent of these folk treatments for cancer in "quack" salad have, in fact, some effect on some type of cancer, I figure I'm improving my odds by indulging my flight into folk fantasy. Genetically, the cancer odds are stacked against me. I quit smoking to change my odds. I eat unhusked sunflower seeds to improve my odds. And I'll eat "quack" salad, hoping to further improve my odds.

It's for sure that my quack salad will contain Vitamins A, C, and E, and some other antioxidents in addition to a lot of fiber. Few people haven't heard about the anticancer activity of these ingredients.

CHAPTER 6

The Spices of Life

Variety may not only be the spice of life, it may be the staff of life. At least, that is my hypothesis. Witch hunters among us comb the literature to find out what poisons are in our food chains. In Phytotoxin Tables (Duke, 1977), I contribute to the witchhunt. Trying to avoid criticism from pesticide advocates, I pointed out that on an LD-50 basis, caffeine, a naturally occurring toxin in many American beverages, is more toxic than many of the herbicides we use. That statement was designed to show that natural toxins can be more toxic than unnatural toxins, especially some of the herbicides used to curb the weeds competing with our crops. Pesticide advocates missed the point and thought I was gunning for 2, 4D, not caffeine. That section of the introduction was editorially muted, even emasculated. On the other hand, environmentalists reviewing the introduction thought I was advocating pesticides.

Some witchhunters look for carcinogens like safrole in sassafras tea, while ignoring the caffeine, etc., we drink. To help correct for my contribution to the herbal witchhunt, I now take the opposite stance. I list some culinary folk remedies for cancer (Hartwell 1967-71) which contain anticancer ingredients. In this sense, I use the same logic used by Farnsworth, Bingel, Fong, Saleh, Christenson, and Saufferer (1976). They surveyed the literature for the distribution in plants of the major oncogenic substances. "Since known oncognic substances are present in these plants, each must be considered as potentially oncogenic, even though direct experimental data are lacking."

Conversely, I superfically checked the literature for the distribution in plants of chemicals cited by Hartwell (1976) or Lewis and Elvin-Lewis (1977) as anticancer substances. Shall I not paraphrase Farnsworth et al to say "Since known antitumor substances are present in these plants, each must be considered for its antitumor potential even though direct experimental data are lacking."?

I confess some skepticism about my approach. I suspect that if any plant species is exhaustively studied, there will be found substances which are oncogenic and antitumor in the sense of the papers presented in the symposium, Plants and Cancer (1976, Cancer Treatment Reports).

Considering this antitumor potential, many herb teas might help in the war against cancer. Maybe so. Maybe not. I do not know. Rutin, which shows up in many spices and herb teas, is said to inhibit tumor formation on mouse skin by the carcinogen benzo(a)pyrene (Leung, 1980). So far rutin has not been reported to cause cancer. I bet I could design an experiment with massive dosages of rutin that would cause cancer in experimental mice.

Rutin was once used in the U.S. to treat capillary hemorrhage due to increased fragility in degenerative vascular diseases (e.g. arteriosclerosis and hypertension), diabetes, and allergic manifestations. If I were suffering from such, I would not hesitate to take a daily quart of herb tea containing anise, buchu, chamomile, coriander, fennel, onion, peppermint, and/or yarrow, all of which are reported to contain rutin.

I do not ask others to drink herbal tea rich in rutin, nor vitamins A, C and E, and even more controversially, amygdalin (closely akin to laetrile).[1] Evidence suggests that many herbal teas are less dangerous than chocolate, coffee, cola, guarana, mate', or tea, the major beverages of all America, all containing caffeine, which Leung (1980) describes as carcinogenic, mutagenic, and teratogenic.

Several other pharmacological properties have been ascribed to rutin, antiedemic, antiatherogenic (in chicks fed cholesterol), increasing survival time in rats on a thrombogenic diet, hypotensive, antiinflammatory, antispasmodic, and protecting against x-radiation. Chinese use the buds of *Sophora japonica* (with ca 20% rutin) for bleeding piles, enterorrhagia, hematuria, hemoptysis, and for preventing strokes.

Many important anticancer compounds also cause cancer. Even tannins (both gallic acid and catechin types) have both carcinogenic and anticancer properties and have been implicated in human cancers (Lewis and Elvin-Lewis, 1977). Certain antitumor agents such as colchicine, demecolcine, podophyllotoxin, and vinblastine are also oncogenic. "Almost all clinically useful antitumor agents, both natural and synthetic, are also carcinogenic" (Farnsworth et al, 1976). Might I not rephrase Farnsworth's sentiments and say "Almost all carcinogenic compounds, both natural and synthetic, should be investigated as potential clincally useful anticancer agents?"

According to Lancet (Oct. 1, 1983), there are ca 300 doctors practicing homeopathy in Great Britain alone. Homeopaths believe a little of an herbal drug will cure what a lot of the same herbal drug will cause. One classical case demonstrating this concept is the "hair of the dog", the drink on New Year's morning to cure the hangover from the excess of New Year's Eve. There are many cases of small doses having the opposite effect of large doses. Alantolactone is hypoglycemic in large doses, hyperglycemic in small doses, e.g. a little bit of a good thing is OK, too much of any compound can be toxic. Might not that little bit brace the immune system for the toxic dose?

According to Lewis and Elvin-Lewis (1977), protein changes known to be induced by chemical carcinogens or oncogenic virus infections in malignant cells are referred to as tumor antigens. The recognition and response of the immune system to these specific carcinogens correspond to the patient's ability to

---

[1] Sadoff et al (1978) report on Rapid Death Associated with Laetrile Ingestion (JAMA 239 No. 15, p. 1532, April 14).

resist cancer. Destruction of malignant cells may be mediated by specific cytotoxic antibodies or sensitized phagocytes. With this mechanism working, malignant transformations, which constantly arise, are constantly eliminated before they reach cell proportions where immunological control is no longer effective (Lewis and Elvin-Lewis, 1977).

I interpret this to mean that a small dose, i.e. something in the range of so dilute a solution as the homeopath's tenth dilution up to the herbalist's tea, of a carcinogen, such as those reported for comfrey (lasiocarpine), gotu kola (asiaticoside), or sassafras (safrole), might trigger the development of antigens for such carcinogens or their biological products. Development of such antigens might better prepare the body for larger insulting doses of the carcinogen than had the body not been exposed to the dilute affront. If there is anything to this, the polyphagous individual, by virtue of wide ranging exposure to small doses of the "spices of life", should be less subject to carcinogenic affront of these spices than the oligophage or monophage.

This wild hypothesis, that variety may be the cancer preventive of life, could be investigated epidemiologically as well as experimentally. How do we test the hypothesis that "variety is the anticancer spice of life"? Provide equal lots of cancer-prone mice with a basic nutrient solution, and ten experimental lots to ten parts each of one herb known to contain an agent that is both oncogenic and antitumor. Expose ten lots to mixtures of one part each of all ten herbs, for the same period of time. Following a few weeks of this regimen, expose all lots to larger doses of pure compound, separated from the herbs. I will take the liberty of suggesting ten reported to contain both carcinogenic and antitumor agents; and all of which have been consumed at one time or another by human beings: celandine, chocolate, coffee, cola, comfrey, cycad, mayapple, mountain ash, poppy, and sassafras.

Moderation seems to contribute to cancer prevention. Further to my variety hypothesis, let me repeat that very small doses may alert the immune system, while large doses may overload the unalerted system. On one point, the director of NIH agrees with me; too much drinking (especially alcohol), eating (especially meat), and/or smoking (especially tobacco) will contribute to one or another type of cancer. On the other hand, moderate habits are encouraged. One hears that two drinks a day might even be beneficial, that smoking risk is proportionate to the number of cigarettes smoked by the smoker or those with whom the smoker selfishly shares his atmosphere, and that both underweight and overweight are to be discouraged. I do not think the Tobacco Lobby would propose studies to see if a very small dose of smoke might, in fact, lower the susceptibility of the poor experimental mouse to the carcinogens of tobacco smoke. It would also surprise me if the Alcohol Lobby would support studies to see if minute traces of alcohol might help prevent some of the ailments caused by large doses of alcohol on the liver. Perhaps the Meat Lobby might support studies to show that minute doses of meat might result in lower incidence of cancer of the colon than exclusive carnivory or exclusive herbivory. But is there any lobby to investigate whether small, not massive, doses of asiaticoside, caffeine, catechin, colchicine, cycasin, demecolcine, dimethylnitrosamine, gallic acid, isatidine, lasiocarpine, macrozamin, methylazoxymethanol, monocrotaline, neocycasin, nitrosonornicotine, parasorbic acid, podophyllotoxin, quinine, reserpine, retronecine, retrorsine, rotenone, safrole, sanguinarine, seneciphylline, shikimic acid, tannin, thymol, and vincaleucoblastine might be beneficial? Is it not remotely possible that minute, even homeopathic, doses of at least some of these might immunize their host against later carcinogenic insults by large doses, accidentally or intentionally ingested?

If so, then once again, there will be circumstantial evidence that eating or drinking small portions of some of the potentially toxic herbal beverages, like basil, chocolate, coffee, cola, guarana, mate, sassafras, or tea, might immunize the drinker to larger doses of carcinogenic chemicals found therein or chemicals with nearly similar structures. If ingestion of a little of a lot of things immunizes the polyphage to more carcinogens than the monophage is exposed to, then epidemeology should show that the rural polyphage has fewer cancers than the rural oligophage. Injecting mice with oncogens or antitumor agents will not test this variety hypothesis. The experimental animals must ingest a variety of herbs and herb teas. For a list of culinary herbs, once regarded as folk remedies for some type of cancer, that have been found to contain some type of anticancer compound, see S in Chapter 9, Herbs and Man. Some of these herbs are relatively safe, some are relatively dangerous. All contain at least one substance listed by Hartwell (1976) as an anticancer agent.

CHAPTER 7

Herbalist's Vitachart

Have you looked at a cereal box or at a nutrition book, trying to figure out what value a food holds for you? When you see that a plant has 18 mg niacin per 100 g (3 ½ oz.), do you know that one 100-g serving of that vegetable satisfies the RDA (recommended dietary allowance) established by the NAS (National Academy of Sciences)? Have you read a lot of articles talking about foods low in one nutrient, high in another, yet not known what they meant by high or low? Such prompted me to publish the Vegetarian Vitachart (Duke, 1977).

In that Vitachart I adopted, for better or worse, a definition of high and low, based on the RDA for the 25 year old male or female, whichever had the higher RDA. High (H) is consistently defined as containing 10% up to 100% of the 25-year old person's RDA in a 100 g serving, raw, as purchased. Low (L) means that 100 g contains 1 to 10% of the RDA. Extremely low (E) means that 100 g contain less than 1% of the RDA. Very high (V) means that 100 g contains as much as or more than 100% of the RDA.

For calcium, a food containing 800 mg or more/100 g is very high (V), a food containing 80 up to 800 mg calcium/100 g is scored high (H), a food containing 8 up to 80 is scored low (L), and a food containing less than 8 mg is scored E (extremely low). These scores all were derived from the edible portion as purchased raw. The amount of calcium in 100 g of any herb scored V for calcium, would satisfy or more than satisfy reference man's RDA for calcium.

A food containing 270 to 2700 calories/100 g was scored high (H), one with 27 up to 270 low (L), and one with less than 27 scored E. After consultation, I accepted 10 g/100g or more as very high (V) for fiber, 1 to 10 g as high (H), and 0.1 to 1 as low (L) and less than 0.1 as extremely low (E). For iron I accepted 18 mg/100 g or more as very high (V), 1.8 to 18 mg/100 g as high (H), 0.18 to 1.8 as low (L), and less than 0.18 as extremely low (E). For magnesium I accepted 350 mg/100 g or more as very high, 35 to 350 as high, 3.5 to 35 as low, and less than 3.5 mg as extremely low. For niacin I accepted 18 mg/100 g or more as very high (V), 1.8 to 18 as high (H), 0.18 to 1.8 as low (L), and less than 0.18 mg as extremely low (E). For phosphorus I accepted 800 mg/100 g or more as very high (V), 80 to 800 as high (H), 8 to 80 as low (L), and less than 8 mg as extremely low (E). For potassium I accepted 2500 mg/100 g or more as very high (V), 250 to 2500 as high (H), 25 to 250 as low (L), and less than 25 as extremely low (E). For protein, I accepted 56 g/100 g or more as very high (V), 5.6 to 56 as high (H), 0.56 to 5.6 as low (L), and less than 0.56 as extremely low (E). For riboflavin, I accepted 1.8 mg/100 g or more as very high (V), 0.18 to 1.8 as high (H), 0.018 up to 0.18 as low (L), and less than 0.018 as extremely low (E). For sodium I accepted 4600 mg/100 g or more as very high (V), 460 to 4600 as high (H), 46 to 460 as low (L), and less than 46 mg as extremely low (E). For thiamin I accepted 1.4 mg/100 g or more as very high (V), 0.14 to 1.4 as high (H), 0.014 to 0.14 as low (L), and less than 0.014 as extremely low (E). For vitamin A I accepted 5000 IU or 3000 ug (micrograms) beta-carotene equivalent/100 g or more as very high (V), 500 to 5000 IU or 300 to 3000 ug carotene as high (H), 50 to 500 IU or 30 to 300 ug carotene as low (L), and less than 50 IU or 300 ug carotene as extremely low (E). For vitamin C I accepted 45 mg ascorbic acid/100 g or more as very high (V), 4.5 to 45 as high (H), 0.45 to 4.5 as low (L), and less than 0.45 mg as extremely low (E). Trace values and zero values were scored (E).

Readers should remember that these figures, taken from the open literature, are for raw materials, some of which may be poisonous. Some poisons are inactivated in cooking, others are not. Many of the vitamins are destroyed by drying, heating, or cooking. This vitachart does not imply that all the listed plants are edible, nor that when they are processed they will still contain the amount of nutrients tabulated. Where different sources gave different nutritional scores, especially if these were given a different common name, multiple entries were made. Without voucher specimens, I accept the taxonomy of the source books, although a few nomenclatural changes have occassionally been made (Duke, 1978). With these reservations in mind, casual readers should find this a quick and handy reference to the nutrient content of dozens of herbs. For comparative purposes, I classify average foods from Latin America in Table 9.

If more and more Americans are consuming more and more herbal teas and salads, in diets that could be either salubrious or deleterious, we need nutritional analyses of the unlisted herbs. I can grow enough of them to provide the pound necessary for nutritional analyses. I cannot however fund the nutritional analyses which I think are seriously needed.

These literal herbal symbols indicate just the "cubby hole" values for the herbs, as purchased (APB) but they are useful for comparison. Real values on the APB basis are presented in Table 11. A forest herb may be 90% water, whereas the dried herbs may be closer to 5 to 10% water. To make comparisons even more valid, I have converted our dozen herbs to the zero-moisture basis (ZMB). This conversion enables even more realistic comparisons. Sure, a raisin is richer in percentage iron than the grape from which it was dried. It didn't gain any iron of course, it lost water. I think the ZMB comparisons are most valid. They are mde by multiplying the unit by $1/100^1$ X where X is the percentage of water.

In general, leafy herbs on a ZMB are rich in calcium and iron, Vitamins A and C, and fiber, most of which are nutritionally imporant. Given the choice I prefer to take my vitamins and minerals, even my protein, from an herb tea or even an herbal dip (like dipping snuff) than from a capsule. Those wishing to cut back on snacks and/or smoking will find that a little dip of the culinary herb, tucked away in a corner of the mouth like a dip of snuff, will curb the whimsical appetite for snacks. I don't know whether it was coincidence or herbal dips that helped me drop ten pounds during the Christmas Season of 1983. While not promoting herbs or vitamins for self-medication, I present a brief tabulation of some of the ailments suspected to respond to vitamin treatments, in Chapter 8.

Table 9. Comparison of Various Types of Latin Foods
(100 grams, as purchased)
Source: Duke, 1977

	Calories	Calcium	Fiber	Iron	Niacin	Phos-phorus	Protein	Ribo-flavin	Thiamin	Vit. A	Vit. C
Nuts & Seeds	H(521)	H(273)	H(36)	H(4.3)	H(5.2)	H(522)	H(16.8)	H(0.28)	H(0.78)	E(17)	L(2.1)
Pulses	H(354)	H(102)	H(5.5)	H(7.1)	H(2.3)	H(398)	H(25.4)	H(0.24)	H(0.58)	E(20)	L(1.9)
Cereals	H(352)	L(74)	H(4.0)	H(4.8)	H(2.7)	H(346)	H(11.7)	H(0.25)	H(0.41)	E(13)	L(0.8)
Vegetables	L(74)	L(26)	H(1.5)	L(1.2)	L(1.0)	L(52)	L(1.8)	L(0.05)	L(0.09)	H(595)	H(31.0)
Fruits	L(93)	L(20)	H(1.4)	L(0.8)	E(0.1)	L(33)	L(1.2)	L(0.06)	L(0.05)	L(35)	H(29.0)
Plant Average (Ave. of 50)	H(279)	L(99)	H(3.2)	H(3.6)	H(2.2)	H(270)	H(11.4)	H(6.18)	H(0.38)	L(135)	H(13.0)
Meat (Ave. of 10)	H(202)	L(19)	E(0.0)	H(3.3)	H(5.9)	H(215)	H(20.6)	H(0.19)	L(0.08)	E(2)	L(1.0)

129

Table 10. Herbalist's Vitachart

Common Name and plant part[1]	Scientific Name[2]	Calcium	Calories	Fiber	Iron	Magnesium	Niacin	Phosphorus	Potassium	Protein	Riboflavin	Sodium	Thiamin	Vitamin A	Vitamin C
acerola (f)	Malpighia punicifolia	L	L	L	L		L	L	L	E	L	E	L	E	V
alfalfa (sh)	Medicago sativa	L	L	H	H		L	L		H	L		L	V	V
allspice (l)	Pimenta dioica	H	L	H	H		L	L	H	L	L		L	V	V
amaranth (l)	Amaranthus tricolor	H	L	H	H		L	L	H	L	L	E	L	V	V
anise (s)*	Pimpinella anisum	H	H	V	V			H	H	H		L			
applemint (1)*	Mentha rotundifolia	V							H	H		E			
basil (l)*	Ocimum basilicum	V	H	V	V		H	H	V	H	H	E	H	V	V
bay (l)*	Laurus nobilis	V	H	V	V		H	H	H	H	H	E	E	V	V
black nightshade (1)	Solanum nigrum	H	L	H	H		L	H		L	H		L	H	H
black pepper (s)*	Piper nigrum	H	L	V	V		L	H	H	H	H	E	L	L	H
bushmint (s)	Hyptis spicigera	H	H	V				H		L			L	L	
caraway (s)*	Carum carvi	H	H	H	H		H	H	H	H	H	E	H		
carob (f)	Ceratonia siliqua	H	L	H				H		L		E		L	
cardamon (s)*	Elettaria cardamomum	H	H	V	H				H	H	H	L	H		
celery (l)*	Apium graveolens	V	E	V	V		L	L	H	L		L		E	H
chard (l)	Beta vulgaris	H	L	L	H	H		H	H	H	H	L	L	L	H
chervil (l)*	Anthriscus cerefolium	V	E	V	V		L		H	L		L	H	V	
chicory (g)	Cichorium intybus	H	L	L	L	L		H	H	L	L		L	L	H
chive (l)	Allium schoenoprasum	L	L	V	L		L	L	H	L	L		L	V	V
clove (f)*	Syzygium aromaticum	H	H	H	L		H	H	H	H	H	L	L	H	V

130

Table 10. Herbalist's Vitachart (cont.)

Common Name and plant part[1]	Scientific Name[2]	Calcium	Calories	Fiber	Iron	Magnesium	Niacin	Phosphorus	Potassium	Protein	Riboflavin	Sodium	Thiamin	Vitamin A	Vitamin C
coriander (s)*	Coriandrum sativum	H	H	V	H		H	H	H	H	H	E	H	E	H
cumin (s)*	Cuminum cyminum	V	H	V	V		L	H	H	H	H	L	H	H	H
curly dock (l)	Rumex crispus	L	E	L	H		L	L		L	L		L	H	V
curly parsley (l)	Petroselinum crispum	H	L	L	H	H		L	H	L	H	L	L	H	H
dandelion (l)	Taraxacum officinale	H	L	H	H		L	H	L	L	H	E	H	V	H
daylily (buds)	Hemerocallis flava	H	L	V	L	L	H	H	H	H	H	E	H	V	V
dill (s)*	Anethum graveolens	V	H	V	V		H	L	V	H	H	L	H	L	
dill (l)*	Anethum graveolens	V	H	V	V	L	L	L	H	L	L	E	L	L	
endive (l)	Cichorium endivia	H	E	L	L	M		H	H	H	H		H	H	H
fennel (l)	Foeniculum vulgare	H	L	L	H		H	L	H	L	H	L	H	H	H
fennel (s)*	Foeniculum vulgare	V	H	V	V	H	L	H		L	L		H	L	L
fenugreek (s)	Trigonella foenum-graecum	H	H	H	V			H	H	H	H	E	H	L	L
fenugreek (l)	Trigonella foenum-graecum	L	L	L			L	L		L	L	L	L	L	
garlic (r)	Allium sativum	L	L	H	L		L	L	H	L	H		L	L	H
ghostplant (l)	Artemisia lactiflora	L	L	H	H	M	L	L	H	L	L		L	H	H
ginger (r)	Zingiber officinale	H	L	H	H			H		H	H		H	E	L
ginseng, American (r)*	Panax quinquefolius	H	L	V	V	H	H	H	H	H	H	E	H	L	H
ginseng, Korean (r)*	Panax schinseng	H		L			L	L		L	L	L	H	V	L
gotu kola (l)	Centella asiatica	H	H	H	H		H	V		L	L		H	E	E
hempseed (s)	Cannabis sativa	H	H	H	H	L	L	L		L	L		H	V	H
high mallow (l)	Malva sylvestris	H	L	V	H		L	L	H	L	H	E	H	V	H
hoary basil (l)	Ocimum canum	H	L	L	H			L		L	H		L	V	V
horseradish (r)	Armoracia rusticana	H	L	H	H		H	L		H	L		L		V
horsetail (l)	Equisetum arvense	L	E	H	H			H	H	H	L	E	E	H	V

131

Table 10. Herbalist's Vitachart (cont.)

Common Name and plant part[1]	Scientific Name[2]	Calcium	Calories	Fiber	Iron	Magnesium	Niacin	Phosphorus	Potassium	Protein	Riboflavin	Sodium	Thiamin	Vitamin A	Vitamin C
lamb's quarter (l)	Chenopodium album	H	L	H	L		L	L		L	H		H	V	V
laurel (bayleaf)	Laurus nobilis	H	L	H	H		L	L	H	L	H	M	L	H	V
leek (l)	Allium ampeloprasum	L	L	H	L		L	L		L	L		L	E	H
lemongrass (l)	Cymbopogon citratus	L	L	H	H		L	L	H	H	L	L	L	H	L
mace (dry)	Myristica fragrans	H	H	V	V		H	H	H	L	H	L	H	H	
marjoram (l)*	Origanum majorana	H	L		H	L	L	L		L	H	L	L	V	V
mint (l)	Mentha sp.	H	L	H	L		L	L	L	L	H	M	H	H	V
mugwort (l)	Artemisia vulgaris	H	H	H	H		H	L	H	H	H		L	H	V
mustard (s)*	Sinapis alba	H	L	H	H	H	H	V		L	H	E	H	L	
mustardgreen (l)	Brassica juncea	H	L	H	H		L	L	H	L	H	M	L	V	V
nightshade (l)	Solanum nigrum	L	H	H	H		L	L	H	H	L	M	L	L	V
nutmeg (dry)	Myristica fragrans	H	E	H	H		L	L		H	L	M	H	H	
onion tops (l)	Allium cepa	V	L		L		L	L	L	L	L	M	L	L	
orange peel	Citrus sinensis	H	L	V	V		L	L	L	H	L	M	L	L	V
oregano (l)*	Origanum vulgare	V	L	H	H	H	L	L	H	L	L	L	H	L	
parsley (l)	Petroselinum crispum			H				L		H		M	L	H	
pennyroyal (l)*	Mentha pulegium	H	E	H	H	L	L	L		E	H	M	L	L	H
peperomia (l)	Peperomia pellucida	L	L	H	L		H	L	L	L		M	H	V	V
pepper, red (f)	Capsicum annuum	L	E	V	L		L	L		L	H	M	L	V	V
pepper, sweet (f)	Capsicum annuum	H	H		H		H	L	H	H	H		L	L	E
perilla (s)	Perilla frutescens	H	L		L		L	L		L	L	M	H	E	H
plantain (l)	Plantago major	L	L	H	L		L	L	L	L			L	H	H
poke (l)	Phytolacca americana	L	E		L		L	L	H	L	H	M	L	V	V
poppy (s)*	Papaver somniferum	V	H		H		L	V	H	H	L		H	V	V

132

Table 10. Herbalist's Vitachart (cont.)

Common Name and plant part[1]	Scientific Name[2]	Calcium	Calories	Fiber	Iron	Magnesium	Niacin	Phosphorus	Potassium	Protein	Riboflavin	Sodium	Thiamin	Vitamin A	Vitamin C
purslane (l)	Portulaca oleracea	H	E	H	H		L	L	H	L	L		L	L	H
rhubarb	Rheum rhaponticum	H	E	L	L	L	L	L	H	L	L	E	L	L	H
rocket (l)	Eruca sativa	H	E	L	L		L	L		L	L		E	L	H
roselle (f)	Hibiscus sabdariffa	H	L	H	E		L	L	H	L		L	H	H	V
rosemary (l)*	Rosmarinus officinalis	V	H	V	V		H	H	H	L	H	L	H	V	H
sage (l)*	Salvia officinalis	V	H	V	V		H	H	H	H		E	H	E	E
savory (l)*	Satureja hortensis	V	H	V	V	H	H	H	H	H	H	E	H	E	E
sesame (s)	Sesamum indicum	V	H	H	H		H	H		H		L	H	E	
sorrel (l)	Rumex acetosa	L	L	L	H		L	L		L	L				H
Spanish thyme (l)	Coleus amboinicus	H	L	L	H		L	L	H	L	H	E	L	H	H
sweet basil (l)	Ocimum basilicum	H	L	H	H		H	L	V	L	H	L	L	V	
tarragon (l)*	Artemisia dracunculus	V	H	V	V		H	H	H	H	H	L	H	H	E
thyme (l)*	Thymus vulgaris	V	H	V	V		L	H	H	H	H		H	H	V
turmeric (r)	Curcuma domestica	L	L	L	V		L	L		L	V	E	L	E	V
watercress (l)	Nasturtium officinale	H	E	L	L	L	L	L		L	H	L	L	H	H
wolfberry (d f)	Lycium chinensis	L	H	V	L		H	H		H	H	L	H	V	V
wolfberry (l)	Lycium chinensis	H	L	H	V	L	L	L		L	V	L	L	V	V
wormseed (l)	Chenopodium ambrosioides	H	L	H	H		L	L		L	H	L	L	H	H

*From Duke and Atchley, 1984, the unmarked entries from Duke, 1977.

[1] d = dry, f = fruit, fl = flower, g = green, l = leaf, m = mature, r = root, s = seed, sh = shoot (or bud).

[2] Authorities for many of these can be found in Duke and Terrell, 1974, Crop Diversification Matrix: Introduction, Taxon 23 (5/6): 759-799.

[3] E = extremely low, L = Low, H = High, V = Very High.

133

Table 11. Proximate Analysis of Herbs.

		Cal	H₂O %	Prot g	Fat g	Total Carb g	Fiber g	Ash g	Ca mg	P mg	Fe mg	Na mg	K mg	Vit A IU	Thia mg	Rib mg	Nia mg	Vit C mg
Basil (ground)	(APB)	251	6.4	14.4	4.0	61.0	17.8	14.3	2123	490	42.0	34	3433	9375	0.15	0.32	6.95	61
	(ZMB)	268	0.0	15.4	4.3	65.2	19.0	15.3	2268	523	44.9	36	3668	10016	0.16	0.34	7.43	65
Caraway (seed)	(APB)	333	9.9	19.8	14.6	49.9	12.6	5.9	689	568	16.2	17	1351	363	0.38	0.38	3.61	—
	(ZMB)	370	0.0	22.0	16.2	55.4	14.0	6.5	765	630	18.0	19	1499	402	0.42	0.42	4.00	—
Chives (leaves)	(APB)	28	91.3	1.8	0.3	5.8	1.1	0.8	69	44	1.7	—	250	5800	0.08	0.13	0.50	56
	(ZMB)	321	0.0	20.7	3.4	66.7	12.6	9.2	793	506	19.5	—	2873	66666	0.92	1.49	5.70	643
Dill (leaves)	(APB)	253	7.3	20.0	4.4	55.8	11.9	12.6	1784	543	48.8	208	3308	—	0.42	0.28	2.81	—
	(ZMB)	273	0.0	21.6	4.7	60.2	12.8	13.6	1924	586	52.6	224	3569	—	0.45	0.30	3.03	—
Fennel (seed)	(APB)	345	8.8	15.8	14.9	52.3	15.7	8.2	1196	487	18.5	88	1694	135	0.41	0.35	6.05	—
	(ZMB)	378	0.0	17.3	16.3	57.3	17.2	9.0	1311	534	20.3	96	1857	148	0.44	0.38	6.63	—
Garlic (powder)	(APB)	332	6.4	16.8	0.8	72.7	1.9	3.3	80	417	2.7	26	1101	Trace	0.47	0.15	0.70	—
	(ZMB)	355	0.0	17.9	0.9	77.7	2.0	3.5	85	446	2.9	28	1176	Trace	0.50	0.16	0.74	—
Marjoram	(APB)	271	7.6	12.7	7.0	60.6	18.1	12.1	1990	306	82.7	77	1522	8068	0.28	0.32	4.12	51
	(ZMB)	293	0.0	13.7	7.6	65.6	19.6	13.1	2154	331	89.5	83	1647	8732	0.30	0.35	4.46	55
Oregano (dried lf)	(APB)	306	7.2	11.0	10.2	64.4	15.0	7.2	1576	200	44.6	15	1669	6903	0.34	—	6.22	—
	(ZMB)	330	0.0	11.9	11.0	69.4	16.2	7.8	1698	216	48.1	16	1738	7439	0.37	—	6.70	—
Parsley (dried)	(APB)	276	9.0	22.4	4.4	51.7	10.3	12.5	1468	351	97.9	452	3805	23340	0.18	1.23	7.93	122
	(ZMB)	303	0.0	24.6	4.8	56.8	11.3	13.7	1613	386	107.6	497	4181	25648	0.20	1.35	8.71	134

APB = As Purchased Basis
ZMB = Zero Moisture Basis

134

Table 11. Proximate Analysis of Herbs (cont.)

		Cal	H₂O %	Prot g	Fat g	Total Carb g	Fiber g	Ash g	Ca mg	P mg	Fe mg	Na mg	K mg	Vit A IU	Thia mg	Rib mg	Nia mg	Vit C mg
Sage (ground)	(APB)	315	8.0	10.6	12.7	60.7	18.0	8.0	1652	91	28.1	11	1070	5900	0.75	0.34	5.72	32
	(ZMB)	342	0.0	11.5	13.8	66.0	19.6	8.7	1796	99	30.5	12	1163	6413	0.82	0.36	6.22	35
Tarragon	(APB)	295	7.7	22.8	7.2	50.2	7.4	12.0	1139	313	32.3	62	3020	4200	0.25	1.34	8.95	—
(ground)	(ZMB)	320	0.0	24.7	7.8	54.4	8.0	13.0	1234	339	35.0	67	3272	4550	0.27	1.45	9.70	—
Thyme	(APB)	276	7.8	9.1	7.4	63.9	18.6	11.8	1890	201	123.6	55	814	3800	0.51	0.40	4.94	—
(ground)	(ZMB)	299	0.0	9.9	8.0	69.3	20.2	12.8	2050	218	134.1	60	883	4121	0.55	0.43	5.36	—

APB = As Purchased Basis
ZMB = Zero Moisture Basis

CHAPTER 8

Herbal Orthomolecularity

According to Linus Pauling, father of orthomolecularity, "orthomolecular" means having the right molecules in the right concentrations," from "ortho" meaning right or correct, like orthodox. "The right molecules are those that are normally present in the human body. Many of them are required for life — such as the vitamins and essential amino acids. Ortho-molecular medicine is the prevention and treatment of disease, the preservation of good health, by varying the concentration of these molecules in the human body". "The powerful drugs that are used by doctors, who treat crises with these crisis drugs, do the job, but they usually have serious side effects, and you have to be careful about them. In particular, they shouldn't be taken day after day, whereas the vitamins can be taken day after day for the rest of your life" (Pauling, 1977). He goes on to cite evidence of success with orthomolecular psychiatry.

Ordinary treatment with phenothiazines for acute schizophrenics has ca 35-45% success. But with "ortho-molecular" treatment in addition to the phenothiazine and whatever else the psychiatrist wants to give them, it is said that 80 percent of them are released and not hospitalized a second time. They are to continue the vitamins the rest of their lives. The phenothiazine they stop taking quickly." (Pauling, 1977).

Perhaps Pauling's orthomolecularity is best known in regards to Vitamin C and the common cold. Williams (1983) says "The most famous — and the most controversial — potential cold preventitive is Vitamin C, long championed by Nobel Laureate Linus Pauling." Pauling theorizes that increasing the intake of Vitamin C strengthens the intercellular cement such that invading viruses are stopped or slowed down.

Many vitamins and minerals are essential. Lack of any one or more of them can cause serious complications. Taking more than the RDA, many times more than the RDA, may not necessarily be a good or safe thing. There are clear signs that large doses of the fat-solubles like Vitamins A and D can be harmful. Without endorsing the high daily regimen that Linus takes himself as reported in The Complete Book of Vitamins, I here repeat them for the readers edification: 1,200 units Vit E, 50 mg thiamine, 50 mg riboflavin, 50 mg pyridoxine, 100 mg niacinamide (and usually also 300-400 mg nicotinic acid), 4000 units Vit A (sometimes 25,000), plus other vitamins and minerals. Ironically in this paragraph he didn't say how much Vitamin C he took. It would be very difficult to acquire such *high* doses via the herbal liqueur, salad or tea. Perhaps that explains the greatest contradiction in Emaus, Pa., home of two of my favorite magazines, *Organic Gardening* and *Prevention*. *Organic Gardening* urges organic matter (manures, mulches, etc.) as sources of minerals and vitamins for plants, but *Prevention* promotes pure pills, rather than whole plants as sources of vitamins and minerals for man. Big farmers usually prefer to apply pure chemical fertilizers right out of the bag. Ironically the orthomolecular believer's daily dosage of vitamins and minerals looks like a chemistry set. It is quicker to pop a vitamin pill than to eat a carrot, just as it is quicker to chemically fertilize a farm than to manure it. I prefer the cabbage-carrot-citrus approach to the pill-pill-pill approach, especially since I also need the fiber and trace minerals and vitamins that occur in herbs. As you saw in Chapter 7, many herbs are rich in vitamins (and fiber). There are few fools like me who eat their herbal tea grounds after they drink their tea, but many of the vitamins, minerals, and almost all of the fiber is there. That's why I also dip, chew, and swallow the more salubrious of the GRAS herbs. It's at least probably better for me that the three packs of cigarettes and ten cups of coffee a day I gave up before switching to herbs.

The roles of vitamins in medicine, especially preventitive medicine, have been long recognized. Lancet (September 10, 1983) notes that 5 million Asian children may develop xerophthalmia due to Vit. A deficiency. Herbal teas and greens could reduce this number significantly. Lately ortho-molecular advocates may have overpromoted so that establishment physicians may label the pharmaceutical purveyors of vitamins as quacks. Hence it is stimulating to read in so prestigious a journal as *Science* (Ames, 1983) that "Vitamin E (tocopherol) is the major radical trap in lipid membranes and has been used clinically in a variety of oxidation-related diseases". Ames lists this first among a variety of small molecules in our diet with antioxidative and possibly anticancer activity. He also lists Vitamins A and C. Since my vitachart doesn't include the all important Vitamin E (tocopherol), I am drawing upon Rodale's COMPLETE BOOK OF VITAMINS to enumerate some of the high tocopherol plants (Gerras et all, 1977). Herbs, especially fresh herbs, are good sources of Vit. A, C and E, all seemingly championed by the orthomolecular establishment. Herbs high in Vitamin A and C are indicated in the Herbalist's Vitachart.

Since our Vitachart Tables are based on 100 g proportion that's what I use here for Vitamin E:

nettle leaf contains	22 IU / 100 g
cabbage leaf (outer)	9 IU / 100 g
mint leaf	7 IU / 100 g
carrot leaf	4 IU / 100 g
dandelion leaf	4 IU / 100 g
nasturtium leaf	4 IU / 100 g
spinach leaf	4 IU / 100 g
asparagus	4 IU / 100 g
broccoli	3 IU / 100 g
ginseng	1.5 IU / 100 g
celery leaf	1 IU / 100 g

Tips for preserving Vitamin E might apply to other vitamins in our herbal teas. If fresh raw leaves are dropped into boiling water rather than brought to a boil, Vitamin E loss is minimized. Gradually bringing a leaf to a boil costs a lot of tocopherol. Vitamin E, the "free-radical scavenger," can more than double the life expectancy of human cells in test tubes. Even the fiber, charted in our Herbalist's Vitachart, is said to have both antioxidant and anticarcinogenic activities.

Nutrition plays a vital role in preventitive medicine and in recuperation. Some orthomolecular theories have proven out, others have not. It is easy to take small doses of Vit. A, C and E in an herb tea, yet not enough to overdose. This may be one of the most beneficial attributes of herb tea, over and above the other therapeutic and placebo potentials. The "medicinal" attributes of nutrients that follow have been extracted largely from Ames, 1983; Bender, 1973; Duke, 1975; Gerras et al, 1977; NAS, 1974).

In Table 12, I present a long list of nutrients, many of which are found in plants, parenthetically indicating plant sources of some of these nutrients. Other sources for some of the more common nutrients were also indicated in Tables 9 and 10.

Table 12. Symbols Used
in Medicinal Applications of Vitamins

Symbol   Name   (Plant Sources)

A   = Vitamin A (green & yellow vegs., esp. carrot)
Ara = Arachidonic Acid (?)

B   = B comples, not specified
B1 = Thiamin (cereals, potatoes, beans, leaves)
B2 = Riboflavin (legumes, leaves, cereals)
B3 = Niacin (nuts, legumes, cereals)
B6 = Pyridoxine (cereals, nuts, legumes, bananas)
B12 = Cyanocobalamin (yeast, comfrey)
B13 = Orotic Acid (?)
B15 = Pangamic Acid (apricot, rice, seeds)
B17 = Laetrile (seeds and leaves of Rosaceae)
Bio = Biotin (cereals, nuts)

C   = Vitamin C (leaves, fruits)
Ca = Calcium (leaves)
Cal = Calories (oilseeds)
Cho = Choline (legumes)
Co = Cobalt
Cr = Chromium
Cu = Copper

D   = Vitamin D (Solanum)

E   = Vitamin E (leaves, cereals, legumes)

F   = Vit. F or Unsat. Fatty Acids
Fat = Cat (oilseeds)
Fe = Iron (leaves)
Fib = Fiber (leaves)
Fl = Fluorine
Fol = Folic Acid (leaves)

GLA = Gamma Linolenic Acid (evening primrose, borage, hops)
Glu = Glutamic Acid (cereals, beets)

I   = Iodine
Ino = Inositol (cereals)

Lec = Lecithin (legumes)
Lin = Linoleic Acid (oilseeds)
Lys = Lysine
K   = Vitamin K (leaves, legumes, potato, tomato)

Table 12. (cont.)

Met = Methionine
Mg = Magnesium
Mn = Manganese (leaves)
Mo = Molybdenum

Na = Sodium (halophytes)
Nia = Niacin

P = Bioflavanoid (Vit. P) (citrus, buckwheat)
Pab = PABA (cereals)
Pan = Pantothenic Acid (seeds)
Ph = Phosphorus
Pot = Potassium
Pro = Protein (legumes, leaves)

S = Sulfur
Se = Selenium (Se-accululators)

T = Vit. T or Sesame Seed Factor (sesame)
Thiamin = Vit. B1
Tocopherol = Vit. E (cereals, leaves)
Try = Tryptophan

U = Vit. U (cabbage)

Zn = Zinc

Somewhere in the Vitamin Literature, the following diseases or ailments are hinted to be helped by the following vitamins (For symbols, see Table 12, p. 141).

Table 13. Disorders Reported to be Corrected with Nutrients

Abortion: E, P
Abruptio placentae: E
Abscess: A, B, C, E
Acne: A, B, B2, B6, C, Ca, D, E, F, Nia, Pan, Pot
Adrenal Exhaustion: B, B2, B12, C, Fol, Na, Pan, Pot
Alcoholism: A, B, B1, B2, B6, *B12**, B15, C, Cho, D, E, Fe, Fol, GLA, K,
          Lec, Mg, Nia, Pan, Pot, *Pro*, Zn
Allergic Rhinitis: (See Hay fever)
Allergies: A, B, B12, C, Ca, D, E, F, Mn, Pan, Pot
Amblyopia: A, B, B1, B12, C, E
Anemia: B, B1, B2, *B6, B12, C,* Ca, Cho, Co, *Cu, E,* Fe, Fol, Met, Mo, Pab,
          Pan, Pro, T
Angina Pectoris: A, B. B12, B15, C, Cho, E, I, Pot, Pro, Se
Anorexia: B1, *Bio,* Nia, *P*
Arteriosclerosis: A, B, B6, C, Ca, Cho, Cr, E, Fol, I, Ino, Mg, *Nia,* P, Ph, Zn
Arthritis: A, B, B2, B6, B12, C, Ca, D, E, F, Fol, I, Lec, Mg, Nia, P, Pan,
          Ph, Pot, Pro, S
Asthma: A, B, B6, B12, B15, C, Cho, D, E, F, Ino, Mn, Pan
Atherosclerosis: A, B, B6, B12, B15, C, Ca, Cho, Cr, E, Fol, I, Ino, Mg, Nia,
          P, Ph, Zn
Athlete's Foot: A, C, E
Autism: B6
Backache: B, C, Ca, D, E, Mg, Nia, Ph, Pro
Baldness: B, B2, B6, Bio, C, Cho, Cu, E, Fol, Ino, Nia, P, Pab, Pan
Bedsores: A, B, B2, C, Cu, D, E, Nia, Pro
Bell's Palsy: B, B1, B6, C, Pro
Beriberi: B, *B1,* C
Birth Defects: B2, B12, Fol
Bitot Spots: A, D, Pro
Boils (Furuncles): A, C, E
Bronchitis: A, C, D, E, F, Pro
Bruises: B, C, D, E, Fe, Fol, K, P
Bruxism (Tooth Grinding): Ca, Pan
Burns: A, B, C, D, E, F, Pab, Pot, *Pro,* Zn
Bursitis: A, B, B12, C, E, Pro
Cancer:A, B, B2, B6, B15, B17, *C,* Cho, D, *E,* Fe, *Fib,* Ino, K, Nia, Pan,
          Ph, Pot, Pro, *Se*
Canker: A, B, C, D, Nia

---

*If italicized, there seems to be acceptable scientific evidence that the vitamin or nutrient does in fact correct or ameliorate the ailment.

Table 13. (cont.)

Carbuncle: A, C, D, E, I
Cardiac Arrest: Cho
Cardiovascular Disease: B1
Caries: *Fl*
Cataracts, A, B, *B2,* C, Ca, D, E, Pan, Pro
Catatonia: Fol
Celiac Disease: A, B, B6, B12, C, Ca, D, E, Fe, Fol, K, Mg, Pro (no gluten)
Cheilosis: *B2, B6*
Chicken Pox: A, C, Pro
Chilblains: D
Chills: C
Cirrhosis (Liver): A, B, B12, B15, C, Cho, D, E, F, Ino, K, Mg, Pro, Zn
Cold: A, B, B6, *C,* Ca, D, E, F, P, Pro
Colitis: A, B6, C, Ca, E, F, Fe, Fib, Mg, Ph, Pot, Pro
Congestive Heart Failure: A, B, B1, E, Pot
Conjunctivitis: A, B, *B2,* C, D, Nia
Constipation: A, B1, C, Ca, Cho, D, E, F, Fat, Fib, Ino, Nia, Pab, Pot
Convulsions: *B6, Mg*
Coronary Thrombosis: *E, F*
Cramps: Ca, E, Pan
Cretinism: I
Croup: A, C, Pro
Cystic Fibrosis: A, B, C, *Cal,* D, E, K, Na, Pro
Cystitis: A, B6, C, D, E, Pan
Dandruff: A, B, B6, E, F
Deafness: A
Dehydration: Na
Dementia: *Nia*
Depression: B1, B2, B6, C, Pab, Pan, Try
Dermatitis: *A, Ara,* B, *B2, B6, Bio,* D, F, *Lin, Nia,* Pot, Pro, S
Desquamative Erythroderma: Bio
Diabetes: A, B, B1, B2, B12, B15, C, Ca, D, E, F, Fe, Mg, Mn, Nia, Pot, Pro, Zn
Diabetes Mellitis: B6
Diabetes Retinitis: P
Diarrhea: A, B, B1, B2, B6, C, Ca, Cl, F, Fe, *Fib,* Fol, Mg, Na, *Nia,* Pan, Pot, Pro
Diathesis: E
Diverticulitis: B, C, *Fib,* Fol
Dizziness: B, B1, B2, B6, B12, C, Ca, Cho, E, Ino, Nia
Dwarfism: *Zn*
Dysmennorrhea: Ca, Fe, GLA
Dyspepsia (Indigestion): B1, B2, B6, Fol, Nia, Pan
Dysplasia: *C*
Eczema: A, Ara, B, B6, Bio, C, Cho, D, E, F, GLA, Ino, S

Table 13. (cont.)

Edema: B, B1, B6, C, Cu, *E,* P, Pot, Pro

Emphysema: A, B, B15, C, D, E, Fol, Pro

Encephalomalaria: *E*

Enteritis: C

Epidermylosia Bullosa: E

Epilepsy: A, B, *B6,* B12, B15, C, Ca, D, E, Fat, Fol, Glu, Mg, Nia, Pan

Eyestrain: A, B, C, D, E

Fainting Spells: Pan

Fatigue: A, B, B2, B12, *Bio,* C, D, *E,* Fe, Fib, Fol, I, Mn, Nia, *P, Pan*

Fever: A, B, B1, C, Ca, *Cal,* D, Na Ph, Pot, Pro

Forgetfulness: B1, E, I, T

Fracture (Bone): A, C, Ca, D, Mg, Pan, Ph, Pot, Pro

Frostbite: Nia

Furuncles (See Boil): A

Gallstones: A, B, C, D, E, K, Pro

Gastritis: A, B, B6, B12, C, D, E, Fe, Fol, Ino, Lec, Lin, Pan, U

Gastroenteritis: A, *B12,* C, *Pan,* Pot

Gingivitis (Gum Disorder): A, B6, C, Ca, D, F, Fe, Fl, Mg, Na, Nia, P,
                           Ph, Pot, Pro

Glaucoma: A, B, B2, C, Cho, D, Ino, P

Glossitis: *B2, B6,* B12, Bio, *Fe, Fol,* Nia

Goiter: A, C, Ca, *I*

Gout: A, B, C, Ca, E, Fe, Mg, Pan, Ph, Pot

Grinding Teeth (See Bruxism): Pan

Gum Disorder (See Gingivitis): A

Hair: A, B, C, I, Pro

Halitosis: A, B, B6, C, Nia

Hangover: B1, B6, C, Cys

Hayfever (Allergis Rhinitis): A, B, C, E, Pan

Headache: A, B, B1, B2, B6, B15, C, Ca, E, *Nia,* Pab, *Pan,* Pot

Heart Disease: B, B1, B6, C, Ca, Cho, E, F, I Ino, Lec, Mg, Nia, Pan, Ph, *Se*

Heavy Metal Toxicity: C

Hemophilia: A, C, Ca, Nia, P, T

Hemorrhage: Bio, C, K, P

Hemorrhoids: A, B, B6, C, Ca, E, P

Hepatitis: A, B2, B15, C, *Cho,* F, K

Hepatosis: Se

Herpes Zoster (Shingles): A, B, B1, B6, B12, C, D

Hodgkin's Disease: Zn

Hypercholesteremia: B, B6, B15, C, Cho, D, E, F, Ino, Lec, Mg, Nia, P, Zn

Hyperexcitability: *Mg*

Hyperoxia: *E*

Hypertension: B, B15, C, Ca, Cho, E, Ino, Lec, Mg, Nia, P, Pot, Pro

Hyperthyroidism: A, B, C, Ca, Cho, I, Ino, Pro

Hypochondriasis: B2

Table 13. (cont.)

Hypoglycemia: B, B6, B12, C, Cho, Cr, F, Fat, Pan, Pro
Hypogonadism: *Zn*
Hypomania: B2
Hypoprothrombinemia: *K*
Hypothroidism: I
Hypoxia: B15, C, *E,* P
Hysteria: B2
Ichthyosis: A
Ileus (Inhibition of Bowel Mobility): Pan
Immune System: B12, Cho, Fol, Met
Impetigo: A, C, D, E
Impotence (Lack of Sex Drive): B6, Zn
Indigestion (See Dyspepsia): B, B1, B6, Fol, Nia, Pan
Infection: A, B, B6, C, Cho, Fol, Pan
Influenza: A, B, B1, B2, B6, C, Nia, Pan, Pro
Insomnia: B, B6, B12, C, Ca, D, Mg, Nia, Pan, Ph, Pot
Intermittent Claudification: E
Itch: Pan
Jaundice: A, B6, C, Ca, D, E, F, K, Lec, Mg, Pan, Ph, Pro, S
Kidney Stones (Renal Calculi): A, B6, C, E, Mg
Kwashiorkor: A, C, Cr, Cu, D, E, Fe, Fol, K, Mg, *Pro,* Se
Leg Cramp: B, B1, B2, B6, Bio, C, Ca, D, E, F, Mg, Na, Pan, Ph, Pro
Leukemia: B, B12, C, Cu, Fe, Fol, P, Zn
Lupus Erythematosus: Pab
Lupus Vulgaris: *D*
Marasmus: *Pro*
Measles: A, C, E, Pro
Megaloblastic Anemia: *Fol*
Megaloblastic Erythopoisis: *B12*
Menieres' Syndrome: B, B1, B2, Ca, E, F, Nia
Meningitis: A, C, D, Pro
Menopause: B6, E
Menstruation: Fe
Mental Illness: B, B1, B6, C, Ca, E, F, *Fol,* Mg, *Nia,* Pan, Ph, Pro
Migraine: Nia
Miscarriage: E
Mononucleosis: A, B1, B2, B6, Bio, C, Cho, Pan, Pot, Pro
Multiple Sclerosis: B, B1, B2, B6, B12, B13, B15, C, Cho, D, E, F, GLA,
             Lec, Mg, Mn, Nia, Pan, Pro
Mumps: C
Muscle Pain: Bio
Muscular Dystrophy: A, B6, B12, C, Cho, E, Ino, Nia, Pan, Pot, Pro, *Se*
Myasthenia Gravis: B1
Myelitis: B12, Pot
Myocardial Infarction: A, B, C, E, Mg, Pot, Pro

Table 13. (cont.)

Nail Ailments: A, C, Ca, Fe, Fol, K, Pro

Nausea: *B6, Bio,* Cl, Pan

Nephritis: A, B, B2, C, Ca, *Cho,* E, Fe, Mg

Nervousness: B6, B12, Mg, Pab

Neuralgia: B12

Neuritis: B, *B1,* B2, B6, *B12, Mg,* Nia, Pan, Pro

Neuropathy: Pan

Neutropenia: *Cu*

Night Blindness: *A,* B, B1, B2, Nia

Nosebleed: P

Obesity: B, B6, B12, C, Ca, E, F, Ino, Lec, Mg, Pro

Odontitis (Tooth Disorder): A, B6, C, Ca, D, F, Fe, Fl, Mg, Na, Nia, P, Ph, Pot, Pro

Osteomalacia: A, C, Ca, D, Mg, Ph

Osteoporosis: B12, C, *Ca,* Cu, D, E, *Fl,* Mg, Ph, Pro

Otitis (Ear Infection): A, C, Pro

Pagetis Disease: *Fl*

Pancreatic Insufficiency: E

Paralysis: Cho

Parkinson's Disease: B, B2, B6, C, Ca, E, Glu, Mg, Nia, Pro

Pellagra: B, B1, B2, B12, Fol, Lys, *Nia,* Pro, Try

Peptic Ulcer (Stomach Ulcer): A, B, B2, B12, C, Ca, E, Fe, Pro, U

Pernicious Anemia: B, B6, *B12,* C, Ca, Co, Cu, E, Fe, Fol, Pro

Phlebitis: B, C, E, Nia, Pan

Pneumonia: A, B, C, D, E, P, Pro

Polio: A, B, C, Na, Pot, Pro

Polymyositis: E

Polyposis: C

Pregnancy: A, B1, B2, B6, B12, C, Cho, D, E, Fe, Fol, Met, Pan, Ph

Prickly Heat: C

Prostatitis: A, B, B6, C, E, F, Pro, *Zn*

Pseudoxanthoma Clasticum: E

Psoriasis: A, B, B2, B6, B12, C, D, E, F, Fol, Ino, Lec, Mg, P, Pan, S

Purpura Hemorrhagica: E

Pyorrhea: A, B, B6, C, Ca, D, Nia, P

Radiation Sickness: B6

Renal Calculi (See Kidney Stone): A

Retarded Growth: B2, Bio, Pan, Ph, Try, Zn

Rheumatic Fever: A, B15, C, D, E, P, Pab

Rheumatism: *B6,* B15, C, Ca, D, E, P, Ph, Pot, Pro

Rhinitis: A, Pro

Rickets: A, C, Ca, D, Mg, Ph

Rickettsia: Pab

Schizophrenia: B1, B6, B12, *C,* Fol, Nia

Sciatica: B, B1, D, E

Table 13. (cont.)

Scleroderma: Pab
Scurvy: A, *C*, Fe, Fol, P, Pro
Seborrhea: *B2, B6,* Bio
Senescence: A, B6, C, Ca, D, *E*, Fe, K
Serum Hepatitis: C
Shingles (Herpes Zoster): A, B, B1, B6, B12, C, D
Sinusitis: A, C, E, Pro
Skin Cancer: C, E, Pab
Sores (External Ulcers): A, B, B2, B12, C, E, Fe, Fol, K, P, Pro
Sore Throat: C
Sprue: E, Nia, Try
Steatorrhea: *E* (Bender, 1973)
Steatosis (of Liver): Cho
Sterility: Ara, *E*, Lin
Stomach Ulcer (See Peptic Ulcer): A, Pot
Stomatitis: *B2, B6,* Nia
Stress: A, *Ara,* B, B1, B2, B6, B12, C, Ca, D, E, Fat, Fol, *Lin,* Nia, Pan,
    Ph, Pot, Pro
Stroke: A, B, C, Cho, E, Ino, Lec, P, Pot, Pro
Sunburn: Ca, E, Pab
Swollen Glands: A, B, C, Pro
Tetany: Ca, D
Thrombophlebitis: *E*
Thrombosis: C, P
Thymus: Choline
Tic: B6
Tingling Feet: Pan
Tonsilitis: C, Fol, Pro
Tooth Decay: B6
Tuberculosis: A, B6, B12, C, Ca, D, Fe, Nia, Pan, Ph, Pro
Ulcer: *U*
Underweight: F, Pro
Vaginitis: A, B, B2, B6, D, E
Varicose Veins: A, B, C, E, P, Pro
Venereal Disease: Pro
Ventricular Arrhythmia: Nia
Vertigo: B2, B12, Nia
Vitiligo: Pab
Warts: A, E
Wernickis Syndrome: *B1*
Worms: A, B, B1, B2, B6, B12, C, Ca, D, F, Fe, K, Pan, Pot, S, Pro
Wounds: C, E, Zn
Wrinkles: C, E
Xerophthalmia: *A*

CHAPTER 9

Herbs and Man [1]

During my first trip to China, "friends" at home changed
the title of a proposed Botanical Society of Washington lecture from "Herbs: Fact and Fiction" to "Herbs and Man".
At first, I was quite comfortable with that, seeing that it
allowed generous latitude. That comfortable feeling was
short-lived. Did my sponsors mean by man the human
species or should it be "huperson" species. Not wanting to
appear sexist or sexless, I decided to alter the title once more
to "Herbs and Man and Woman". That certainly might at-
tract a different audience than a lecture entitled "Herbs and
Sex".

On the other hand, reading between the lines, I believe
that many taxpayer questions I receive are prompted by sexual urges of one kind or another. Ginseng is the herb I am

---

[1]Modified from an illustrated and animated lecture delivered to Botanical Society of
Washington, November 14, 1978.

most inquired about. Many ginseng inquiries are generated by the belief that ginseng makes old men young again. I'm not sure that it helps at all, but there are a lot of younger and older people who think so. Now I find this title Herbs and Man useful as a summary concluding chapter to my Culinary Herbs. Rather than try to cover all uses of all herbs, I've taken an amateurish A to Z approach to a simplified summary of herbal use and abuse. This is largely derived from the herbal literature, not all of which is reliable. In my summary survey, I list those herbs or ingredients stated to be used for this or that. I believe some of them, and have serious doubts about others. Although I have experimented with many herbal teas, I can't really endorse them to others. I know people allergic to peanuts, soybeans, and I myself may be allergic to such common items as asparagus, barley, shrimp, turkey, tomatoes, even wheat and other grains containing gluten. Friends of mine may be immune to these, and allergic, possibly fatally allergic, to others including new food and beverage sources. Let me reiterate here what should accompany any lecture on herbs and/or herbal teas. I've survived thirty years of abusing myself with up to ten cups of coffee, glasses of tea, and/or bottles of beer a day. I don't know which herbs are less toxic than these everyday beverages; I don't know which are slightly more toxic. I don't know which exotic foods, medicinal and herbs are inocuous, which are toxic. I have only an educated guess. Let the buyer beware.

ANAPHRODISIAC
Aconite
Belladonna
Catnip
Chamomile
Chaste tree
Coriander
Cucumber
Hemp
Henbane
Hops
Lettuce
Marjoram
Rue
Sage
Sweetflag
Waterlily

APHRODISIACS
Anise
Basil
Caraway
Cardinal Flower
Celery
Cicely
Coriander
Cottonseed
Cowhage
Damiana
Endive
Fenugreek
Galanga
Garlic
Ginger
Ginseng
Gotu kola
Greenbrier
Harmal
Hemp

APHRODISIACS cont.
Jimsonweed
Joe-pye-weed
Kava
Ladyslipper
Lettuce
Lovage
Mandrake
Muskokra
Nutmeg
Onion
Passion flower
Pepper
Saffron
Sarsaparilla
Savory
Sweetflag
Watercress
Vervain
Withania
Yohimbe

FOR IMPOTENCE
Aristolochia
Burra gookeroo
Eleuthero
False Unicorn root
Ginseng
Jimsonweed

FOR INFERTILE
WOMEN
Arrach

TO DELAY
EJACULATION
Poppy

BALDNESS
Agave
Arnica
Beets
Burdock
Calendula
Garlic
Ginseng
Juniper
Lavender
Nettle
Onion
Rosemary
Sage
Southernwood
Sweetflag
Walnut
Watercress
Yarrow

DANDRUFF
Agave
Beets
Chamomile
Eucalyptus
Horsemint
Olive
Rosemary
Spearmint
Walnut

DEPILATORY
Caper spurge

FACE PACKS
(in clay, curd,
  honey, yogurt, etc.)
Blackberry
Cowslip
Daisy
Dandelion
Elder
Fennel

FACE PACKS cont.
Horsetail
Linseed
Nettle
Sage
Tansy
Yarrow

FACIAL STEAMS
Basil
Burnet
Chamomile
Cornflower
Elder
Fennel
Lavender
Lime
Marigold
Mullein
Nasturtium
Nettle
Peppermint
Sage
Yarrow

FRECKLES
Avens
Burdock
Centaury
Elder
Fumitory
Honeysuckle
Horseradish
Liverwort
Onion
Papaya
Pennyroyal
Rice
Solomon's seal
Spinach
Strawberry

HAIR RINSE
Chamomile
Lemon verbena
Rosemary
Sage

HERBAL BATHS
Basil
Bay
Calamus
Comfrey
Geranium
Lavender
Lovage
Mint
Orris
Rose
Rosemary
Sandlewood
Thyme
Valerian

HERBAL SOAPS
Balsam pear
Batis
Bouncing bet
Entada
Horsetail
Pepperbush
Poke
Prickly poppy
Soapberry
Soapwort
Yucca

SLIMMING HERBS
Aconite
Ash
Balm
Burdock
Calamus
Chickweed
Clivers
Coca

SLIMMING HERBS cont.
Cola
Fennel
Iris
Nettle
Parsley
Turmeric

STIMULATE HAIR
GROWTH
Garlic
Mustard
Nettle
Peach
Rosemary
Southernwood

VARICOSE
VEINS
Avens
Betony
Burnet
Calendula
Cayenne
Chile
Coltsfoot
Deadnettle
Germander
Horsetail
Lavender
Marigold
Marjoram
Melilot
Quinine
Rice
Sage
Sassafras
Shepherd's Purse
Witchhazel

WRINKLES
Comfrey

AMYGDALIN SOURCES

Almond
Apple
Apricot
Bird cherry
Cherry
Cherry laurel
Cotoneaster
Loquat
Mountain ash
Peach
Photinia
Plum
Prune
Quince

CANCER 'CAUSES'[1]

Asiaticoside
Colchicine
Cycasin
Demecolcine
Dimethylnitrosamine
Hydroxysenkirkine
Isatidine
Jacobine
Lasiocarpine
Macrozamin
Methylazoxymethanol
Monocrotaline
Neocycasin A, B, C, E
N-nitrosonornicotine
Parasorbic acid
Podophyllotoxin
Reserpine
Retronecine
Safrole
Sanguinarine
Seneciphylline
Shikimic acid
Vincaleukoblastine

CANCER 'CURES'[1]

Acer Saponin P *(Acer)*
Bruceanthin *(Brucea)*
Camptothecin *(Camptotheca, Mappia)*
Cesalin *(Caesalpinia)*
3-Desmethylcolchicine *(Colchicum)*
Ellipticine *(Bleekeria, Ochrosia)*
Emetine *(Cephaelis)*
Fagaronine *(Fagara)*
Harringtonine *(Cephalotaxus)*
Holacanthone *(Holacantha)*
Homoharringtonine *(Cephalotaxus)*
Indicine-N-oxide *(Heliotropium)*
Lapachol *(Avicennia, Stereospermum, Tabebuia, Tecoma)*
Maytansine *(Maytenus, Putterlickia)*
9-Methoxyellipticine *(Bleekeria, Ochrosia)*
Nitidine *(Fagara)*
Taxol *(Taxus)*
Thalicarpine *(Thalictrum)*
Tripdiolide *(Tripterygium)*
Triptolide *(Tripterygium)*
Tylocrebrine *(Tylophora)*

See also Q for Quackery
See also S for "Spices of Life"

---

[1] Compounds rather than species listed. Homeopaths might argue that small doses of compounds that cause cancer might cure or prevent cancer.

BARK CLOTH
Brosimum
Castilla
Cecropia
Ficus
Hibiscus
Muntingia
Poulsenia
Pseudolmedia

BLACK
Alder
Baneberry
Bramble
Bugle
Cascarilla
Cashew
Dogbane
Elder
Gipsyweed
Poisonivy
Sourwood
Sumac
Walnut
Yellowflag

BLUE
Blueberry
Carrot
Chicory
Cornflower
Elderberry
Elecampane
Greenbrier
Indigo
Larkspur
Sorrel
Woad
Yellowflag

BROWN
Alder
Apple

BROWN cont.
Barberry
Bayberry
Beech
Birch
Black cherry
Bloodroot
Box
Butternut
Chamomile
China
Cockleburr
Coffee
Dogbane
Fustic (young)
Goldenrod
Hemlock
Hickory
Hops
Maple
Marigold
Mountain laurel
Mountain mint
New Jersey tea
Osage orange
Pear
Pecan
Privet
Sassafras
Sourwood
St. John's wort
Sumach
Sunflower
Tea
Teasel
Tobacco
Walnut
Wild marjoram
Willow

FIBER
Abaca
Aramina

FIBER cont.
Baobob
Basswood
Cattail
China jute
Cotton
Ensete
Esparto
Flax
Hemp
Henequen
Jute
Kapok
Kenaf
Milkweed
Nettle
New Zealand flax
Pineapple
Ramie
Redbud
Roselle
Sisal
Snakeplant
Spanish broom
Sunn hemp

GRAY
Bearberry
Blueberry
Bramble
Butternut
Horsetail
Maple
Quack grass
Sumac
Yarrow

GREEN
Alder
Bromegrass
Broom
Cedar
Chamomile

GREEN cont.

Cicely
Coneflower
Elder
Hollyhock
Hyssop
Iris
Lily-of-the-Valley
Mistletoe
Morning glory
Nettle
Oak
Parsley
Plantain
Ragweed
Ragwort
Smartweed
Sourwood
Spinach
Tansy
Walnut
Yarrow
Zinnia

ORANGE

Fustic (young)
Henna
Madder
Safflower

PURPLE

Black cherry
Briar
Cockleburr
Dandelion
Elderberry
Gooseberry
Grape
Oak
Poke
Redcedar
Spindle tree
Sundew
Wild marjoram

RED

Balsam
Barberry
Bedstraw
Beet
Bergamot
Bloodroot
Cardinal flower
Clivers
Cranberry
Dogwood
Hollyhock
Madder
Poke
Poppy
Quinsywort
Safflower
Saunders
St. John'swort
Sorrel

VIOLET

Aloe
Madder

YELLOW

Agrimony
Apple
Aster
Bayberry
Bedstraw
Betony
Birch
Bloodroot
Broom
Buckthorn
Catnip
Cherry
China
Coneflower
Dandelion

YELLOW cont.

Daphne
Elder
Everlasting
Fumitory
Fustic
Goldenrod
Goldenseal
Goldthread
Gorse
Greenweed
Groundsel
Jewelweed
Madder
Marigold
Mountain
Mullein
Onion
Peach
Pomegranat
Pot marigold
Privet
Saffron
Sassafras
Smartweed
Smilax
Southernwood
Spindle tree
Sumac
Sunflower
Sweetgale
Tamarind
Tickweed
Toadflax
Tomato
Turmeric
Wheat
Wild carrot
Woundwort
Yarrow
Zinnia

HALLUCINOGEN
Catnip
Cereus
Coca
Foxglove
Henbane
Jimsonweed
Juniper
Kavakava
Mandrake
Nutmeg
Periwinkle

HERBAL HIGHS
Betel
Broom
Calea
California poppy
Camphor
Catnip
Coleus
Colorines
Damiana
Dona ana
Harmal
Henbane
Hops
Hydrangea
Jimsonweed
Kava-kava
Kola
Lobelia
Mandrake
Mescal bean
Morningglory
Nightshade
Nutmeg
Peyote
Pipizintaintli
San pedro
Valerian
Wild lettuce
Woodrose
Yohimbe

HYPNOTIC
Corkwood
Henbane
Poppy

INSANITY "REMEDIES"
Basil
Bermudagrass
Betel
Borage
Catmint
Catnip
Cotton
Digitalis
Gourd
Hellebore
Hemlock
Hemp
Henbane
Huisache
Jimsonweed
Mandrake
Nettle
Paeony
Paris
Pennyroyal
Plantain
Sowthistle
Strawberry

INTOXICANT
Cascarilla
Clary
Darnel
Dogbane
Henbane
Kava
Linden
Mandioca

RESTRICTED HERBS
Aconite
Belladonna
Calamus
Coca
Cohoba
Cottonroot
Digitalis
Elm
Gelsemium
Harmal
Hellebore
Henbane
Lily-of-the-Valley
Lobelia
Mandrake
Marijuana
Nux-vomica
Opium poppy
Pennyroyal
Peyote
Poke
Savin
Stramonium
Tansy

STIMULANT
Allspice
Aloe
Angelica
Angostura
Arnica
Asafetida
Asarabacca
Basil
Bayberry
Beech
Benzoin
Betel
Birthwort
Boldo
Boneset
Buchu

STIMULANT cont.

Cajuput
Caraway
Cardamon
Carrot
Catmint
Cat thyme
Celery
Cinnamon
Clove
Coca
Coffee
Copaiba
Coriander
Cornflower
Corn
Cowhage
Cumin
Damiana
Dill
Elecampane
Eucalyptus
Fennel
Galbanum
Garlic
Ginger
Ginseng
Gotu kola
Gravelroot
Groundivy
Horseradish
Hyssop
Juniper
Kava
Kola

STIMULANT cont.

Larch
Lavender
Lavender cotton
Lemon
Lovage
Mace
Magnolia
Marigold
Masterwort
Mayapple
Mugwort
Myrrh
Nutmeg
Orange
Pennyroyal
Poppy
Rosemary
Rue
Sage
Sassafras
Selfheal
Snakeroot
Southernwood
Spearmint
Staranise
Sweetflag
Tansy
Tea
Valerian
Wintergreen
Wormseed
Yarrow
Yerba santa
Yellow bugle

FUMITORY

Absinth
Angelica
Anise
Bearberry
Beech
Beet
Belladonna
Betony
Bogbean
Broom
California poppy
Carob
Cascarilla
Catalpa
Catnip
Chamomile
Chile
Cicely
Cinnamon
Coltsfoot
Coriander
Cornmint
Cornsilk
Cubeb
Damiana
Deers-tongue
Eleuthero
Eyebright
Fig
Filbert
Fumitory
Ginseng
Harmal
Hemp
Henbane
Hops
Hydrangea
Jimsonweed
Juniper
Khat

FUMITORY cont.

Licorice
Lobelia
Lovage
Marjoram
Melilot
Mint
Mullein
Mustard
Nigella
Nutmeg
Papaya
Passionflower
Periwinkle
Poppy
Rabbit
Raspberry
Rosemary
Roshagrass
Rue
Sage
Sassafras
Snakeroot
Sumach
Sunflower
Sweetclover
Sweetflag
Thyme
Tobacco
Waxberry
Wild lettuce
Woodruff
Wormwood
Yarrow
Yerba santa
Yohimbe

QUIT OPIUM
Combretum

QUIT TOBACCO
Lobelia

SNUFF[1]
Basil
Germander
Goldenseal
Yarrow

STERNUTATORY
Asarabacca
Bayberry
Bloodroot
Buttercup
Cat thyme
Daphne
Groundivy
Hazlewort
Ipecac
Lavender
Lily-of-the-Valley
Marigold
Mullein
Orris
Pellitory
Poke
Sabadilla
Sassy bark
Soapwort
Solomon's seal
Sweet marjoram
Tobacco
White hellebore

[1]Recently, I lost ten pounds by dipping herbs, i.e., holding them in a corner of my mouth when I was hungry and/or sleepy. Any of the tea herbs worked well with me as a stimulant deappetizer.

CHILDBIRTH

Avens
Basil
Camphor
Chamomile
Cohosh
Comfrey
Cotton
Elder
Flax
Hemp
Horehound
Honeysuckle
Nettle
Olive
Orange
Oregano
Pennyroyal
Raspberry
Rue
Wormwood

CONTRACEPTIVE

Lithospermum
Pea
Rosary pea
Stevia

DYSMENORRHEA

Anise
Asparagus
Balm
Belladonna
Calendula
Catnip
Chamomile
Chicory
Cowslip
Cotton
Elecampane
Goldenrod
Hops
Lovage
Marjoram

DYSMENORRHEA cont.

Mugwort
Parsley
Poke
Rue
Safflower
Saffron
Sage
Sassafras
Tamarind
Thyme
Woodruff
Wormseed

ECBOLIC

Blackroot
Calotropis
Cottonroot
Hemlock
Laurel
Pennyroyal
Red Cedar
Savine
Spurge laurel
Sweetgale
Tansy
Yellow cedar

EMMENAGOGUE

Angelica
Arrach
Bugle
Carrot
Catmint
Cohosh
Costmary
Feverfew
Groundsel
Henna
Motherwort
Mugwort
Parsley
Pennyroyal
Rue

EMMENAGOGUE cont.
Saffron
Sage
Southernwood
Tansy
Wintergreen

HYSTERIA
Absinth
Asafetida
Balm
Basil
Burdock
Camphor
Caraway
Catnip
Chamomile
Clary
Garlic
Gentian
Hellebore
Hops
Horehound
Hyssop
Lavender
Mistletoe
Mugwort
Oregano
Passionflower
Pennyroyal
Peppermint
Poppy
Rue
Saffron
Sage
Tansy
Valerian
Wormseed

LABOR
Basil
Catnip
Cotton
Skunk cabbage

VAGINITIS
Amaranth
Calendula
Fenugreek
Goldenseal
Horsetail
Myrrh
Ragwort
Tansy
Thyme
Walnut
Wintergreen
Witchhazel
Yarrow

HALITOSIS

CHEW STICKS
Acacia
Alder
Birch
Citrus
Coffee
Cola
Dogwood
Mango
Mangosteen
Miracle fruit
Neem
Poplar
Sassafras
Sweetgum
Tamarind
Walnut
Wintergreen

HALITOSIS
Agrimony
Angelica
Anise
Avens
Basil
Calamus
Caraway
Cardamon
Clove
Coriander
Cress
Dill
Fennel
Horsemint
Horsetail
Mace
Mastic
Mint
Muskseed
Nutmeg
Orris
Parsley
Pennyroyal

HALITOSIS cont.
Peppermint
Rosemary
Rosinweed
Rue
Sage
Spearmint
Staranise
Thyme

GARGLE
Acacia
Agrimony
Bistort
Borage
Cudweed
Groundivy
Honeysuckle
Lemon
Loosestrife
Lovage
New Jersey Tea
Nightshade
Pellitory
Periwinkle
Pomegranate
Poppy
Pyrola
Ragwort
Rhatany
Rose
Sage
Sanicle
Selfheal
Silverweed
Sorrel
Spearmint
Stonecrop
Stoneroot
Thyme
Tormentil
White mustard

MASTICATORY
Balata
Chicle
Coca
Dogbane
Fig
Milkweed
Peach
Rosinweed
Spurge
Sweetgum
Tobacco

MOUTHWASH
Bayberry
Cudweed
Goldenseal
Myrrh
Nightshade
Pondlily
Sage
Sassafras
Sea lavender
Thyme

THRUSH
Airpotato
Cladonia
Dasheen
Fig
Garlic
Goldthread
Grape
Houseleek
Indian sarsaparilla
Lime
Myrrh
Persimmon
Plantain
Raspberry
Sage
Selfheal

ANIMAL ALLIES OF PERFUMES
Ambergris
Castoreum
Civet
Musk

HERBAL CANDLES
Clove
Lavender
Lemon
Mint
Rose
Rosemary
Sage
Thyme

INCENSE (BURN)
Balm
Cascarilla
Cinnamon
Cloves
Frankincense
Lavender
Lovage
Marjoram
Orris
Rosemary
Rue
Sage
Santolina
Southernwood
Star anise
Tansy
Thyme
Vetiver

PERFUMES
Angelica
Balsam
Benzoin
Bergamot
Carnation

PERFUMES cont.
Cassie
Cinnamon
Citronella
Frankincense
Galbanum
Gardenia
Geranium
Hyacinth
Iris
Jasmine
Jonquil
Labdanum
Lavender
Lemon
Muskseed
Myrrh
Narcissus
Neroli
Olibanum
Orange
Patchouly
Quince
Rose
Rosemary
Sandal
Storax
Tonka
Tuberose
Vanilla
Verbena
Vetiver
Violet
Ylang ylang

POTPOURRIS
Allspice (spice)
Angelica
Anise
Balm
Basil
Borage
Calamus (fix)

POTPOURRIS cont.
Cardamon
Cassie (oil)
Cinnamon (spice)
Clove (spice)
Coriander
Jasmin (oil)
Lavender
Lemon
Lemon verbena
Lovage
Mace
Marjoram
Mullein
Neroli (oil)
Nutmeg (spice)
Orange
Orangemint
Orris (fix)
Patchouli (oil)
Rose (fix) (oil)
Rosemary
Sage
Sandalwood
Sweetclover (oil)
Tarragon
Thyme
Vanilla
Vetiver

STREWING
Balm
Basil
Chamomile
Costmary
Fennel
Germander
Hyssop
Lavender
Marjoram
Mint

STREWING cont.
Pennyroyal
Sage
Savory
Tansy
Thyme

TOILET WATER
Angelica
Balm
Chamomile
Cinnamon
Clove
Fennel
Horsetail
Lavender
Lemon
Lemon verbena
Lovage (a deodorant)
Mint
Nutmeg
Pennyroyal
Peppermint
Pine
Rosemary
Sage
Thyme
Valerian

TUSSIE-MUSSIE
Lavender
Lemonthyme
Lemonverbena
Marigold
Marjoram
Peppermint
  geranium
Pineapplesage
Rose geranium
Rosemary
Sage
Thyme

JELLIES

ASPICS
Bay
Dandelion
Dill
Fennel
Marjoram
Oregano

**CANDIED
FLOWERS**
Anchusa
Bergamot
Borage
Chrysanthemum
Dandelion
Lavender
Marigold
Nasturtium
Rose
Rosemary
Sage

CONSERVES
Balm
Borage
Cicely
Lavender
Lovage
Mint
Rose
Rosemary
Spearmint
Sweetflag
Violet

JAMS
Angelica
Balm
Mint
Rhubarb
Rose
Rosemary

JELLIES
Balm
Cardamon
Geranium
Horehound
Lavender
Lemon
Mint
Parsley
Peppermint
Rosemary
Sage
Spearmint
Tarragon
Thyme

K KITCHEN K

BUTTER
Basil
Chive
Dill
Garlic
Horseradish
Marjoram
Mint
Oregano
Parsley
Rosemary
Sage
Savory
Tarragon
Thyme

DIPS
Basil
Chive
Cumin
Dill
Marjoram
Onion
Oregano
Parsley
Rosemary
Savory
Thyme

HERB MUSTARD
Basil
Celeryseed
Chives
Marjoram
Oregano
Savory
Thyme

HERB PEPPER
Allspice
Cinnamon
Marjoram
Pepper

HERB PEPPER cont.
Rosemary
Savory
Thyme

HERB SALTS
Basil
Bay
Cayenne
Celeryseed
Chive
Clove
Dill
Fennel
Garlic
Marjoram
Nutmeg
Oregano
Parsley
Rosemary
Sage
Tarragon
Thyme

KITCHEN WINDOW GARDEN
Anise
Balm
Basil
Chive
Coriander
Dill
Mint
Oregano
Parsley
Rosemary
Sage
Savory
Tarragon
Thyme
Watercress

MEAT
TENDERIZER
Fig
Papaya
Pineapple

OMELETS
Basil
Chive
Dill
Fennel
Pepper
Savory
Tansy
Tarragon

SALAD
Borage
Chard
Chervil
Chicory
Comfrey
Cress
Dandelion
Fenugreek
Mustard
Purslane
Sorrel
Watercress

VINEGARS
Basil
Burnet
Chervil
Dill
Fennel
Garlic
Lovage
Marjoram
Mint
Oregano
Parsley
Rosemary
Sage
Savory
Shallot
Spearmint
Tarragon
Thyme

CHARTREUSE ADDITIVES

Angelica
Balm
Basil
Cicely
Coriander
Hyssop
Orangemint
Peppermint
Sweetflag
Tansy

HANGOVER "CURES"

Bugle
Dill
Gelsemium
Ginger
Ginseng
Hazelwort
Lemon
Marjoram
Mother-of-thyme
Parsley
Tabasco
Thyme

LIQUEURS

Aloe
Angelica
Angostura
Anise
Balm
Caraway
Cicely
Cinnamon
Clove
Coriander
Cumin
Elecampane
Fennel
Hyssop

LIQUEURS cont.

Lavender
Licorice
Nutmeg
Orris
Peppermint
Requieni mint
Rue
Sandalwood
Tansy
Thyme
Tonka
Woodruff
Wormwood
SEE ALSO CHAP 4

NATURAL COLORINGS

Annatto (pink-orange)
Beet (red)
Bergamot (red)
Cochineal
Fennel (green)
Grape (red)
Marigold (orange)
Paprika (red)
Poke (purple)
Potmarigold (orange)
Safflower (orange)
Saffron (yellow)
Turmeric (yellow)

"SOBERUPPERS"

Banana
Citron
Coffee
Ivy
Lemon
Loquat
Onion
Orange
Quinine

VERMOUTH ADDITIVES

Angelica
Balm
Caraway
Celery
Chamomile
Cinnamon
Cloves
Dill
Elecampane
Fennel
Gentian
Ginger
Nutmeg
Peppermint
Rosemary
Sage
Spearmint
Staranise
Sweetflag
Thyme
Wintergreen
Wormwood
Yarrow

MEDICINE
Proven Pharmacological Phytochemicals[1]

Flowering Plants That Currently (1980) Are Sources of Drugs Useful in the United States
(Farnsworth and Loub, 1983)

Plant Names	Family	Type of Drug
Ammi majus	Umbelliferae	*Xanthotoxin*[a]
Ananas comosus	Bromellaceae	Bromelain
Atropa belladonna	Solanaceae	Belladonna Extract
Avena sativa	Gramineae	Oatmeal Concentrate
Capsicum species	Solanaceae	Capsicum Oleoresin
Carica papaya	Caricaceae	Papain
Cassia acutifolia	Leguminosae	Sennosides A + B
Cassia angustifolia	Leguminosae	Sennosides A + B
Catharanthus roseus	Apocynaceae	*Leurocristine* (vincristine)
		Vincaleukoblastine (vinblastine)
Cinchona species	Rubiaceae	*Quinine*
		Quininidine
Citrus limon	Rutaceae	Pectin
Colchicum actumnale	Liliacae	*Colchicine*
Digitalis lanata	Scrophulariaceae	*Digoxin*
		Lanatoside C
		Acetyligitoxin
Digitalis purpurea	Scrophulariaceae	*Digitoxin*
		Digitalis whole leaf
Dioscorea species (several)	Dioscoreaceae	*Diosgenin*
Duboisia myoporoides	Solanaceae	*Atropine*
		Hyoscyamine
		Scopolamine
Ephedra sinica	Ephedraceae	**Ephedrine*
		*Pseudoephedrine
Glycine max	Leguminosae	Sitosterols
Papaver somniferum	Papaveraceae	Opium
		Codeine
		Morphine
		Noscapine
		*Papaverine

[a]Italicized names indicate single chemical compounds of known structure.
*Also produced by synthesis.

Plant Names	Family	Type of Drug
Physostigma venenosum	Leguminosae	*Physostigmine* (Eserine)
Pilocarpus jaborandi	Rutaceae	*Pilocarpine*
Plantago species	Plantaginaceae	Psyllium husks
Podophyllum peltatum	Berberidaceae	Podophyllin
Prunus domestica	Rosaceae	Prune Concentrate
Rauvolfia serpentina	Apocynaceae	*Reserpine*
		Alseroxylon Fraction
		Powdered whole root Rauwolfia
Rauvolfia vomitoria	Apocynaceae	*Deserpidine*
		Reserpine
		Rescinnamine
Rhamnus purshiana	Rhamnaceae	Cascara Bark
		Casanthranol
Rheum species	Polygonaceae	Rhubarb Root
Ricinus communis	Euphorbiaceae	Castor Oil
		Ricinoleic Acid
Veratrum viride	Liliaceae	*Veratrum viride* Extract
		Cryptennamine

NARCOTIC
(Plants That A Hippy, Reading Grieve's Modern Herbal, Might Assume were Narcotic)[1]

Acontinum napellus
Aesculus hippocastanum
Ajuga reptans
Aletris farinosa
Anamirta paniculata
Andira inermis
Arbutus unedo
Artemisia absinthum
Atropa belladonna
Buxus sempervirens
Cannabis sativa
Cereus grandiflorus
Chelidonium majus
Chionanthus virginicus
Cinnamomum camphora
Conium maculatum
Coriandrum sativum
Croton eleuteria
Cytisus scoparius
Datura arborea
Datura fastuosa
Datura metel
Datura stramonium
Dipteryx odorata
Duboisia myoporoides
Erythroxylum coca
Euphorbia hypericifolia
Ferula foetida
Ferula sumbul
Gelsemium sempervirens
Genista tinctoria
Gleditsia triacanthos
Helleborus niger
Humulus lupulus
Hydrocotyle asiatica
Hyoscyamus niger
Kalmia latifolia
Lachnanthes tinctoria
Lactuca sativa

Lactuca scariosa
Lactuca virosa
Laurus nobilis
Lavandula latifolia
Lavandula stoechas
Ledum latifolium
Ledum palustre
Lobelia inflata
Lolium temulentum
Lophophora lewinii
Lycopus virginicus
Mandragora officinalis
Melampyrum pratense
Melia azadirachta
Myrica cerifera
Narcissus sp.
Oenanthe phellandrium
Papaver rhoeas
Papaver somniferum
Paris quadrifolia
Passiflora incarnata
Passiflora quadrangularis
Passiflora rubra
Paullinia cupana
Phaseolus vulgaris
Physalis somnifera
Phytolacca americana
Picraena excelsa
Pilocarpus jaborandi
Piper methysticum
Piscidia erythrina
Prunus laurocerasus
Ruta graveolens
Salvia sclarea
Scopolia carniolica
Solanum dulcamara
Solanum nigrum
Stachys betonica
Symplocarpus foetidus

---

[1]Back in the old days, narcotic often meant poisonous.

Tilia europaea
Veratrum album
Verbascum thapsus
Viscum album

NARCOTIC ALKALOIDS
atropine
bulbocapnine
cocaine
codeine
corycavine
corydaline
corydine
helleborin
hyoscyamine
morphine
scopolamine
solanine
sparteine
thebaine

BORDERS
Basil
Betony
Bugle
Calamint
Chive
Dead nettle
Parsley
Pennyroyal
Primrose
Santolina
Savory
Thyme
Woodruff

BONSAI
Bay
Juniper
Myrtle
Rosemary
Thyme
Wormwood

HANGING POTS
Cleavers
Gotu Kola
Mint
Oregano
Parsley
Pimpernel
Purslane
Rosemary
Savory
Thyme
Watercress

WREATHS
Annual Wormwood
Balsam
Bugle
Holly
Horehound
Hyssop
Rue
Sage
Santolina
Savory
Thyme
Wormwood

P

PESTICIDE

P

FLEA REPELLENTS
Cedar
Chamomile
Pennyroyal
Pyrethrum
Rue

FUNGICIDE
Chamomile
Dill
Mint
Mustard
Sage
Sumac
Walnut

HERBICIDES
Barley
Bearberry
Crabgrass
Eucalyptus
Fir
Garlic
Hackberry
Jerusalem artichoke
Johnson grass
Oat
Onion
Peanut
Pine
Rye
Spurge
Sunflower
Sycamore
Tobacco
Walnut
Wheat

HERPECIDE
Cherimolia
Coltsfoot
Flax
Gourd

HERPECIDE cont.
Grape
Nightshade
Poisonivy
Poppy
Walnut

INSECT REPELLENT
Absinth
Anise
Basil
Bergamot
Bugle
Cajuput
Caraway
Catnip
Chamomile
Chives
Clover
Coriander
Fennel
Fenugreek
Garlic
Geranium
Hemp
Hot pepper
Hyssop
Lavender
Lemongrass
Marigold
Mayapple
Mint
Nasturtium
Nigella
Onion
Parsley
Pennyroyal
Peppermint
Pyrethrum
Rosemary
Rue
Sage

INSECT REPELLENT cont.
Santolina
Shoofly
Southernwood
Spearmint
Tansy
Thyme
Walnut
Woodruff
Wormwood

INSECTICIDE
Bugle
Cedron
Garlic
Gorse
Horehound
Juniper
Larkspur
Lavender
Lavendercotton
Melilot
Mugwort
Pennyroyal
Rue
Santolina
Sassafras
Smartweed
Southernwood
Sweetflag
Sweetgale
Tephrosia
Toadflax
Tobacco
Tomato
Wormwood
Zinnia

174

MOTH REPELLENT
Lavender
Mint
Rosemary
Santolina
Southernwood
Tansy
Thyme
Wormwood

PEDICULICIDE
Balsam of Peru
Bugle
Cocculus
Henbane
Labrador tea
Larkspur
Lavender
Neem
Sabadilla
Virginia creeper
White hellebore

RINGWORM
Araroba
Basil
Bloodroot
Borage
Burdock
Cashew
Castor

RINGWORM cont.
Celandine
Coltsfoot
Elm
Fig
Garlic
Gourd
Jewelweed
Mugwort
Mullein
Nightshade
Plantain
Poisonivy
Poke
Poppy
Senna
Sorrel
Thyme
Walnut

VIRICIDE
Grape
Lemonbalm
Marigold
Purslane
Radish
Sassafras
Skullcap
(See also Herpicide)(p)
(See also Vermifuge)(v)

QUACKERY
Herbs Once Used as Folk Remedies
for Cancer

QUACK BREAD
Barley
Caraway
Cassava
Corn
Flaxseed
Job'stear
Millet
Poppyseed
Rye
Sesame
Sorghum
Wheat

QUACK BROTH
Alfalfa
Chicory
Cress
Dill
Garlic
Leek
Nettle
Onion
Shepherd's purse
Spinach
Yarrow

QUACK DRESSING
Basil
Black pepper
Coconut oil
Dill
Garlic
Hot pepper
Lemon
Lime
Oil Palm
Olive
Oregano

QUACK FRUIT
COCKTAIL
Akebia
Banana
Cantaloupe
Date
Fig
Mayapple
Papaya
Persimmon
Pineapple
Prickly pear
Rhubarb (petiole)
Watermelon

QUACK FRUIT
JUICE
Barberry
Bilimbi
Cochineal
Cranberry
Durian
Emblic
Litchi
Longan
Maracuja
Mulberry

QUACK
FUMITORIES
Angelica
Anise
Betony
Chamomile
Coriander
Ginseng
Jimsonweed
Lobelia
Melilot
Mullein

QUACK HIGH
Absinth
Angel's trumpet
Catnip
Hellebore
Hemp
Henbane
Hops
Jimsonweed
Lobelia
Nutmeg
Parsley
Poppy
Sage
Valerian

QUACK LIQUEUR
Alkanna
Angelica
Anise
Calamus
Calendula
Cashew apple
Cicely
Clove
Cola
Fennel
Ginseng
Juniper
Licorice
Orris
Saffron
Southernwood

QUACK NUT
BOWL
Acorn
Butternut
Cashew
Chestnut
Hazel

QUACK NUT BOWL cont.

Hempseed
Ginkgo
Indian almond
Jojoba
Pistacho
Safflower
Walnut

QUACK POISONS

Belladonna
Colchicum
Digitalis
Dogbane
Hellebore
Mandrake
Mezereon
Mistletoe
Nightshade
Oleander
Periwinkle
Squill
Tobacco
Woody nightshade
Yew

QUACK POTHERBS

Arnica
Asparagus
Balsampear
Basella
Beet
Cabbage
Comfrey
Orach
Pigweed
Purslane
Sundew
Yarrow

QUACK SALAD

Absinthe[1]
Ammi
Arnica[1]
Atriplex[1]
Beet[1]
Black walnut[1]
Borage[1]
Burdock
Cabbage
Calendula[1]
Celery[1]
Chervil
Chickweed
Chicory[1]
Chive[1]
Chufa[1]
Colocynth[1]
Comfrey
Crimson clover[1]
Crown vetch[1]
Cucumber[1]
Cumin[1]
Elecampane
Flax[1]
Garlic[1]
Gromwell
Hot pepper[1]
Husk-cherry
Lettuce
Licorice[1]
Mousear
Nasturtium
New Zealand spinach
Onion[1]
Orach
Peanut[1]
Poke[1]
Radish
Rampion

QUACK SALAD cont.

Rocket
Safflower[1]
Salvia[1]
Sorrel
Sow thistle
Spinach
Stinging nettle[1]
Tamarind[1]
Tansy[1]
Tea[1]
Tomato[1]
Watercress

QUACK SALVE

Aloe
Balsam
Benzoin
Breadfruit
Camphor
Caper spurge
Cassia
Castor
Cinnamon
Fig
Frankincense
Gum arabic
Jojoba
Myrrh
Storax
Sweetgum
Turpentine
Witchhazel

QUACK SAUCE

Caper
Cayenne
Coriander
Cumin

---

[1] From Duke, 1976. I suggested to McDonalds that they might profit from the introduction of a McQuack Soup & Salad Quick Lunch. They weren't interested. Interested sponsors are invited to contact Jim Duke, Chief Quack, Herbal Vineyard, Fulton, MD 20759.

QUACK SAUCE cont.
Horseradish
Hot pepper
Mustard
Oregano
Soy
Tamarind

QUACK SOUP
Basil
Bay
Bean
Cabbage
Carrot
Chile
Chickpea
Hot pepper
Lentil
Limabean
Nutgrass
Okra
Onion
Parsnip
Pigeon pea
Potato
Pumpkin
Tomato
Turnip
Yam

QUACK TEA
Balm
Betony
Birchbark
Bugle
Bugloss
Calamint
Chamomile
Costmary
Dandelion
Dittany
Fenugreek
Germander
Gotu kola
Herb Robert
Horehound
Horsetail
Hyssop
Jasmine
Lavender
Lovage
Marjoram
Marshmallow
Melilot
Mullein
Pennyroyal
Peppermint
Plantain
Psyllium
Roselle
Rosemary
Sarsaparilla
Sassafras
Skullcap
Tansy
Tea
Thyme
Watermint

QUACK VEGETABLE COCKTAIL
Beet
Brusselsprouts
Burdock
Celery
Chufa
Eggplant
Fababean
Globe artichoke
Malanga
Onion
Pea
Spinach
Squash
Sunflower
Sweet potato
Taro
Turnip

QUACK WINE
Agave
Angelica
Clary
Coltsfoot
Dandelion
Elderberry
Grape
Hops
Labdanum
Mulberry
Rhubarb

TO COAGULATE
MILK[1]
Artichoke
Bedstraw (yellow)
Cheeseberry
Ginger
Nettle
Sorrel
Sundew
Thistle

TO COLOR
CHEESE
Annato
Marigold
Pot marigold
Safflower
Saffron

TO FLAVOR
CHEESE
Basil
Burnet
Caraway
Cayenne
Chervil
Chile
Chive
Cumin
Dill
Lavender
Lovage
Marigold
Marjoram
Mint
Nasturtium
Paprika
Parsley
Poppyseed
Rosemary
Rue
Sage
Tansy
Tarragon
Thyme
Watercress

---

[1] Scientific names of some lesser known rennets include Carduus, Cirsium, Crotalaria, Leucas, Pinguicula, Rhazya, Solanum, Streblus, and Withania.

Achillea millefolium (Yarrow)
Acorus calamus (Sweetflag)
Aloe barbadensis (Aloe)
Allium cepa (Onion)
Allium sativum (Garlic)
Anethum graveolens (Dill)
Angelica archangelica
  (Angelica)
Arctium lappa (Burdock)
Arnica montana (Arnica)
Artemisia dracunculus
  (Tarragon)
Asarum canadensis
  (Wild ginger)
Asparagus officinalis
  (Asparagus)
Berberis vulgaris (Barberry)
Camellia sinensis (Tea)
Centaurium umbellatum
  (Centaury)
Ceratonia siliqua (Carob)
Chamaemelum nobile
  (Chamomile)
Chelidonium majus (Celandine)
Chimaphila umbellata
  (Pipsissewa)
Coriandrum sativum
  (Coriander)
Drimys winteri (Winterbark)
Echinacea angustifolia
  (Coneflower)
Eucalyptus (Eucalyptus)
Eupatorium perfoliatum
  (Boneset)
Gentiana lutea (Gentian)
Glycyrrhiza glabra (Licorice)
Heliotropium indicum
  (Heliotrope)

Hyssopus officinalis
  (Hyssop)
Inula helenium (Elecampane)
Juniperus communis (Juniper)
Laurus nobilis (Bay)
Lavandula latifolia
  (Spike lavender)
Levisticum officinale (Lovage)
Marrubium vulgare
  (Horehound)
Medicago sativa (Alfalfa)
Melissa officinalis
  (Lemonbalm)
Nardostachys jatamansi
  (Indian Valerian)
Origanum majorana (Marjoram)
Origanum vulgare (Oregano)
Panax ginseng (Ginseng)
Pimpinella anisum (Anise)
Prunus serotina (Wildcherry)
Rheum officinale (Rhubarb)
Rosmarinus officinalis
  (Rosemary)
Sassafras albidum (Sassafras)
Satureja montana
  (Winter Savory)
Saussurea lappa (Costus)
Swertia chirata (Chirata)
Symphytum officinale
  (Comfrey)
Taraxacum officinale
  (Dandelion)
Thymus vulgaris (Thyme)
Trifolium pratense (Red Clover)
Valeriana officinalis
  (Valerian)
Zea mays (Cornsilk)
SEE CHAP. 6

CAFFEINE CONTAINERS
Chocolate
Coffee
Cola
Gurana
Mate
Tea

COFFEE SUBSTITUTES
(usually scorched)
Carrot
Cassie
Chicory
Chickpea
Chocolate
Coconut
Dandelion
Holly
Oats
Okra
Wheat
Yaupon

LEMON FLAVOR
Citronella
Citrus
Lemon basil
Lemon balm
Lemon eucalyptus
Lemon geranium
Lemon grass
Lemon pittosporum
Lemon thyme
Lemon verbena
Lime geranium

"PARAMEDICINAL TEAS"[1]

For Bad Appetite
Anise
Avens
Clary
Coriander
Costmary
Fenugreek
Germander
Ginseng
Lavender
Rosemary
Savory
Tarragon
Thyme

For Bronchitis
Angelica
Balm
Betony
Borage
Catnip
Colt'sfoot
Comfrey
Ginseng
Groundivy
Horehound
Licorice
Lovage
Rosemary
Sage
Southernwood
Thyme

For Colds
Absinth
Balm
Basil
Bay
Catnip

"PARAMEDICINAL TEAS"[1] cont.

For Colds
Chamomile
Fennel
Germander
Ginseng
Horehound
Hyssop
Lemon
Licorice
Lime
Mint
Pennyroyal
Peppermint
Rosemary
Sage
Savory
Thyme
Turmeric
Wintergreen
Yarrow

For Diarrhea
Avens
Basil
Burnet
Calendula
Catnip
Colt'sfoot
Comfrey
Hyssop
Peppermint
Sage
Savory
Strawberry
Thyme
SEE CHAP. 3

---

[1]This table constitutes a survey of the literature and should not be construed as prescription. He who self-medicates has a fool for a doctor. Many of these herbs, like many foods, are suspected of containing toxins and carcinogens.

"PARAMEDICINAL TEAS"[1] cont.

For Gout
Angelica
Balm
Basil
Bugle
Burdock
Burnet
Cherry
Cicely
Elecampane
Fennel
Gentian
Germander
Ginseng
Groundivy
Lavender
Marjoram
Pennyroyal
Periploca
Rosemary
Sassafras
Watercress

For Headache
Angelica
Balm
Basil
Catnip
Cumin
Fennel
Germander
Groundivy
Hops
Horehound
Lavender
Marigold
Marjoram
Pennyroyal
Peppermint

"PARAMEDICINAL TEAS"[1] cont.

For Headache
Rosemary
Rue
Sage
Savory
Skullcap
Thyme
Wintergreen
Woodruff
Wormseed

For Indigestion
Anise
Caraway
Catnip
Chamomile
Cicely
Clary
Coriander
Cumin
Dill
Elecampane
Fennel
Ginseng
Goldenseal
Horehound
Hyssop
Juniper
Lavender
Lovage
Marjoram
Mint
Oregano
Parsley
Pennyroyal
Peppermint
Rosemary
Sage
Savory

"PARAMEDICINAL TEAS"[1] cont.

For Indigestion
Spearmint
Thyme
Yarrow

For Nerves
Balm
Basil
Belladonna
Catnip
Chamomile
Dittany
Ginseng
Horehound
Hyssop
Lavender
Marjoram
Oregano
Pennyroyal
Peppermint
Periwinkle
Rosemary
Sage
Savory
Skullcap
Spearmint
Tansy
Thyme
Woodruff
Wormseed

For Rheumatism
Absinth
Angelica
Basil
Boneset
Borage
Buchu
Bugle
Comfrey

---

[1]This table constitutes a survey of the literature and should not be construed as prescription. He who self-medicates has a fool for a doctor. Many of these herbs, like many foods, are suspected of containing toxins and carcinogens.

"PARAMEDICINAL
TEAS"[1]cont.
For Rheumatism
Coriander
Dittany
Fennel
Ginseng
Gotukola
Hyssop
Juniper
Lavender
Marjoram
Melilot
Motherwort
Mugwort
Nettle
Peppermint
Pipsissewa
Rosemary
Rue
Sage
Skullcap
Thyme

'TRAVELER'S TEA'
Basil
Burnet
Celery
Dill
Fennel
Lovage
Marjoram
Mugwort
Nasturtium
Rosemary
Sage
Tarragon
Thyme

SWEETENERS
Angelica
Artichoke
Balm
Cardamon
Chicory
Cicely
Licorice
Miracle fruit
Perilla
Skirret
Stevia
Sugarleaf Hydrangea
SEE ALSO CHAP. 3

---

[1]This table constitutes a survey of the literature and should not be construed as prescription. He who self-medicates has a fool for a doctor. Many of these herbs, like many foods, are suspected of containing toxins and carcinogens.

CYSTITIS
Bearberry
Buchu
Cornsilk
Cubebs
Dogbane
Henbane
Horsetail
Pipsissewa
Quackgrass
Watermelon
Wintergreen
Yew

DIURETIC
Aconite
Agrimony
Arnica
Asparagus
Bearberry
Belladonna
Blueflag
Borage
Bugle
Burdock
Carrot
Celery
Chamomile
Cicely
Coffee
Cornsilk
Damiana
Dandelion
Elder
Elecampane
Fennel
Fenugreek
Foxglove
Fumitory
Garlic
Gotukola
Groundivy
Henbane

DIURETIC cont.
Hops
Horseradish
Jewelweed
Lovage
Lupine
Nightshade
Onion
Orris
Parsley
Passionflower
Peach
Pimpernel
Pipsissewa
Pyrola
Smartweed
Snakeroot
Sorrel
Spearmint
Staranise
Stoneroot
Strawberry
Sumach
Sunflower
Toadflax
Tobacco
Valerian
Wild lettuce
Wintergreen
Wormwood

GONORRHEA
Basil
Boldo
Buchu
Burdock
Cayenne
Chile
Cranberry
Cress
Dandelion
Dogbane
Fenugreek

GONORRHEA cont.
Hemp
Indigo
Juniper
Lime
Nettle
New Jersey tea
Peanut
Pepper
Pipsissewa
Purslane
Radish
Sassafras
Strawberry
Sumac
Tree-of-heaven
Wintergreen
Yarrow

GRAVEL
Absinth
Asparagus
Bearberry
Carrot
Coffee
Dandelion
Horsetail
Hydrangea
Lovage
Marshmallow
Onion
Parsley
Radish
Smartweed
Sorrel
Spearmint
Wormwood

PROSTATITIS
Buchu
Corn
Cubebs
Cucumber

PROSTATITIS cont.

Garlic
Horsetail
Nettle
Parsley
Pumpkin
Quackgrass
Rosemary
Soybean
Squash
Sunflower
Watermelon
Wintergreen

SYPHILIS

Airpotato
Aloe
Benoil
Bermudagrass
Betel
Blueflag
Box
Burdock
Butternut
Clematis
Elder
Fenugreek
Gotukola
Indigo
Lemon
Mountain laurel
New Jersey tea
Nightshade
Onion
Poke
Radish
Rape
Rhododendron
Sarsaparilla
Sassafras
Savine
Spurge
Sumac
Tamarind

SYPHILIS cont.

Turmeric
Walnut

URINARY TRACT

Absinth
Asparagus
Barley
Basil
Bearberry
Bistort
Boldo
Buchu
Burdock
Clivers
Coltsfoot
Cranberry
Cumin
Dandelion
Dodder
Dogbane
Flax
Garlic
Goldenseal
Hemp
Horsetail
Hyssop
Juniper
Licorice
Liverwort
Lovage
Marshmallow
Mate
Matico
Mistletoe
Nightshade
Radish
Sassafras
Senna
Sorrel
St. John'swort
Sweetflag
Turmeric
Wintergreen

VERMIFUGE

HOOKWORM
Caraway
Fennel
Thyme
Wintergreen
Wormseed

MAGGOTS
Tobacco
Water agrimony

RINGWORM
Araroba
Basil
Bloodroot
Borage
Burdock
Cashew
Castor
Celandine
Colt'sfoot
Elm
Fig
Garlic
Gourd
Jewelweed
Mugwort
Mullein
Nightshade
Plantain
Poke
Poisonivy
Poppy
Senna
Sorrel
Sumac
Thyme
Walnut

ROUNDWORM
Hyssop
Wormseed

SALMONELLA
Thyme

TAENIFUGE
Araroba
Areca
Butternut
Mullein
Pomegranate
Tree-of-heaven
Walnut

VERMIFUGE
Alder
Aloe
Avens
Betony
Boneset
Calamint
Carrot
Castor
Catnip
Chamomile
Costmary
Dittany
Elecampane
Garlic
Germander
Horehound
Horseradish
Lavender
Motherwort
Mugwort
Lemon
Melilot
Onion
Papaya
Parsley
Peach
Pennyroyal
Peppermint
Sage
Selfheal
Southernwood
Tansy
Thyme
Walnut
Woodruff
Wormseed
Yarrow

WITCHCRAFT

EXORCISM
Anise
Ash
Basil
Dill
Fumitory
Honesty
Hyssop
Jimsonweed
Mandrake
Mint
Mullein
Peony
Periwinkle
Purslane
Rue
Sage
St. John'swort
Snapdragon
Valerian
Vervain

DEATH
DIVINERS
Gilly-flower
Honesty
Rhubarb
Rosemary
Sage

WITCHCRAFT
Angelica
Apple
Basil
Bay
Black poppy
Calabar
Clover
Coriander
Crocus
Dill
Elder
Elecampane
Fennel

WITCHCRAFT cont.
Fig
Foxglove
Hemlock
Henbane
Hyssop
Garlic
Groundivy
Honesty
Lavender
Milfoil
Mustard
Parsley
Rosemary
Rue
Sandalwood
Snapdragon
Southernwood
Tansy
Thyme
Yarrow

WITCHES
SPELLS
Chervil
Dittany
Hemlock
Henbane
Mandrake
Pennyroyal
Poppies
Vervain

ANTIDOTES
Anise
Angelica
Calabar bean
Elder
Flax
Mignonette
Peony
Pimpernel
Valerian
Wormwood

REPUTED REFRIGERANTS[1]

Catmint
Chickweed
Currant
Houseleek
Lemon
Lime
Plantain
Pomegranate
Rice
Sorrel
Sow thistle
Sumac
Sweetgale
Tamarind
Wild lettuce

WATERPROOFING (RUBBER)

Balata
Dandelion
Fig
Goldenrod
Guayule
Madrono
Manihot
Rubber
Spurge

WATER SOURCE[2]

Bamboo
Cacao
Ceiba
Cuipo
Grape
Lemongrass
Palms

WAX SOURCES

Ash
Banana
Bayberry
Candelilla
Cinnamon
Coconut
Fig
Madrono
Privet
Safflower
Spurge
Sugarcane
Wax palm

---

[1]Reduce thirst and lower fever.
[2]Some tropical genera furnishing water substitutes include Bambusa, Casuarina, Cavanillesia, Ceiba, Cocos, Coffea, Combretum, Costus, Cymbopogon, Davilla, Desmoncus, Doliocarpus, Heliconic, Raphanus, Ravenala, Spondias, Tetracera, Theobroma, and Vitis.

Y YOUTH Y

ANTIOXIDANT COMPOUNDS

Ascorbic acid
Caffeic Acid
Capsicoside
Carotene
Epigallocatechin
Ferulic Acid
Glutathione
Maltol
Nordihydroguairetic Acid
Osajin
Pomiferin
Quercetin
Rosmanol
Tocopherol
Tricin
Uric Acid[1]

ANTIOXIDANT HERBS

Alfalfa
Allspice
Benzoin
Black Pepper
Cabbage
Caraway
Celery seed
Chile
Cinnamon
Clove
Cumin
Fennel

ANTIOXIDANT HERBS cont.

Garlic
Ginger
Lemon
Lemongrass
Mace
Marjoram
Mint
Nutmeg
Olibanum
Pepper
Rosemary
Sage
Savory
Slippery Elm
Stinging Nettle
Tansy
Tumeric

LONGEVITY

Angelica
Coriander
Fennel
Ginseng
Goldenseal
Gotukola
Lemonbalm
Milkweed
Sage
Sundew
Tansy

---

[1]If uric acid is antioxidant and antioxidants reduce cancer incidence as Ames (1983) suggests, then those 8% of Americans who suffer goit should show an epidemiologically lower incidence of cancer. Science should investigate!

Z z z   SLEEP   z z Z

FOR INSOMNIA
Agrimony
Anise
Asparagus
Balm
Belladonna
Bugle
Catnip
Celery
Chamomile
Chicory
Clivers
Dandelion
Deadnettle
Dill
Dogbane
Fumitory
Gelsemium
Ginseng
Grape
Hemlock
Hemp
Henbane
Hops
Jimsonweed
Lemon verbena
Lettuce
Lobelia
Lovage
Mandrake
Marjoram
Melilot
Mugwort
Mullein
New Jersey tea
Nightshade
Oat
Oregano
Peach
Pennyroyal
Peppermint
Poppy
Primrose

FOR INSOMNIA cont.
Rose
Rosemary
Sage
Skullcap
Spearmint
Thyme
Tobacco
Valerian
Witchhazel
Woodruff

FOR MELANCHOLY
Balm
Borage
Bugloss
Burnet
Calamint
Celery
Deadnettle
Dodder
Feverfew
Fumitory
Garlic
Hellebore
Hemp
Lavender
Mandrake
Meadowsweet
Motherwort
Peony
Peppertree
Pimpernel
Pineapple
Saffron
Strawberry
Thyme
Woodruff
Yarrow

Z Z z   SLEEP (cont.)   z Z Z

FOR NIGHTMARES
Catnip
Chamomile
Horehound
Peony
Periwinkle
Rue
Thyme

FOR NIGHTSWEAT
Balm
Goldenseal
Hops
Hyssop
Nettle
Oxeye Daisy
Sage
Strawberry
Walnut

HERB PILLOWS
Angelica
Balm
Bergamot
Dill
Lavender
Lemon verbena
Marjoram
Mint
Peppermint
Rose geranium
Rosemary
Sage
Tarragon
Thyme
Valerian
Woodruff

NOSEGAYS
(Sniffed to keep
awake during
boring sermons)
Bergamot
Cabbage Rose
Coriander
Costmary
Dill
Fennel
Marigold
Moss Rose
Southernwood

"SLEEPING PILLOWS"
(Placed under pillow
to make one sleep)
Agrimony
Alchemilla

References

Ames, B.N. 1983. Dietary Carcinogens and Anticarcinogens. Science 221: 1256-1264.

Avard, L., Story, G., and Wentworth-Jackson, A. 1982. Six Herbs — Development of a Commerical Programme. Australian Hort. (Aug): 93-105.

Bender, A.E. 1973. Nutrition and Dietetic Foods. Chemical Publishing Co., Inc., New York, 298 pp.

Conrow, R., and Hecksel, A. 1983. Herbal Pathfinders. Voices of the Herb Renaissance. Woodbridge Press, Santa Barbara, 286 pp.

C.S.I.R. 1948-76. The Wealth of India. New Delhi, 11 volumes.

Daisley, G. 1982. The Illustrated Book of Herbs. American Nature Society Press, New York, 128 pp.

Duke, J.A. 1975. Crop Chemistry and Folk Medicine, pp. 83-117 in V.C. Runeckles, ed., Advances in Phytochemistry, Vol. 9. Plenum Press, New York, 309 pp.

Duke, J.A. 1976. Quack Salad, pp. 249-251 in Bricklin, M., ed. The Practical Encyclopedia of Natural Healing, Rodale Press, Emaus.

Duke, J.A. 1977. Phytotoxin Tables, Critical Reviews in Toxicology 5(3): 189-237.

Duke, J.A. 1978. The Quest for Tolerant Germplasm. ASA Special Symposium 32.

Duke, J.A. 1981. Handbook of Legumes of World Economic Importance. Plenum Press, New York 345 pp.

Duke, J.A. 1982. Herbs as a Small Farm Enterprise and the Value of Aromatic Plants as Economic Intercrops, pp. 76-83 in Research for Small Farms, Proceedings of the Special Symposium, USDA, Beltsville, MD, 301 pp.

Duke, J.A. 1983. Medicinal Plants of the Bible. Trado-Medic Books, Conch Magazine, Ltd., Buffalo, N.Y., 233 pp.

Duke, J.A. 1984. in ed. Borderline Herbs, CRC Press, Boca Raton. (Newly Titled: Handbook of Medicinal Herbs)

Duke, J.A. and Atchley, A.A. 1984, in ed. Proximate Analysis Tables. CRC Press, Boca Raton.

Duke, J.A. and Wain, K. 1981. Medicinal Plants of the World. A computer index. Unpublished, 3 vols, 88,000 entries.

Farnsworth, N.R., A.S. Bingel, H.H.S. Fong, A.A. Saleh, G.M. Christenson, and S.M. Saufferer. 1976. Oncogenic and Tumor-Promoting Spermatophytes and Pteridophytes and Their Active Principles. Cancer Treatment Reports 60(8): 1171-1214.

Farnsworth, N.R. and Loub, W.D. 1983. Information Gathering and Data Bases that are Pertinent to the Development of Plant Derived Drugs, pp. 178-195 *in* OTA, Plants — The Potentials for Extracting Protein, Medicines and Other Useful Chemicals. Office of Technology Assessment, Washington, DC. 252 pp.

Farrel, 1974. Making Cordials and Liqueurs at Home. Harper & Row.

Foster, G.B. 1973. Herbs for Every Garden. E.P. Dutton & Co.

Fox, H.M. 1933. Gardening with Herbs for Flavor and Fragrance. Reprint 1970. Dover Publications, New York, 334 pp.

Gerras, D., Golant, J. and Hanna, E.J. 1977. The Complete Book of Vitamins. Rodale Press, Emaus, Pa., 814 pp.

Greenhalgh, P. 1979. The Market for Culinary Herbs. Tropical Products Institute, London, 167 pp.

Grieve, M. 1931. A Modern Herbal (reprint 1974) Hafner Press, MacMillan Publishing Co., Inc. New York, 915 pp.

Hallgarten, P.A. 1979. Spirits and Liqueurs. Faber & Faber, London. 176 pp.

Harrop, R, ed. 1977. Encyclopedia of Herbs. Chartwell Books, Secaucus Avenue, N.J. 153 pp.

Hartwell, J.L. 1967-71. Plants Used against Cancer: A Survey. Lloydia 30-34, 11 installments. Issued in One Volume by Quarterman Publications, Lawrence, Mass. 1982.

Hartwell, J.L. 1976. Types of Anticancer Agents Isolated from Plants. Cancer Treatment Reports 60(8): 1031-1067.

Heinerman, J. 1979. Science of Herbal Medicine. Bi-World Publishers, Orem, Utah, 318 pp.

Hylton, W.H. ed. 1974. The Rodale Herb Book, Rodale Press Book Division, Emaus, Pa. 653 pp.

Kirschmann, J.D. 1975. Nutrition Almanac. McGraw-Hill, New York, 263 pp.

Leung, A.Y. 1980. Encyclopedia of Common Natural Ingredients. Wiley-Interscience, John Wiley & Sons, New York, 409 pp.

Lewis, W.H. and Elvin-Lewis, M.P.F. 1977. Medical Botany. John Wiley & Sons, New York.

Leyel, C.F. 1950. Herbal Delights. Faber & Faber, Ltd., London.

Lowman, M.S. 1946. Savory Herbs: Culture and Use. USDA Farmers Bull. 1977, Washington, D.C. 33 pp.

Martin, J.A., Senn, T.L., Kingman, A., and Ezen, D.O. 1973. (reprint). Herbs for the Home Garden, 34 pp.

Merck, & Co. 1968. The Merck Index. 8th Ed., Merck & Co., Rahway, N.Y. 1713 pp.

Meyer, C. 1977. 50 Years of The Herbalist Almanac, Meyerbooks, Glenwood, Ill. 256 pp.

Morton, J.F. 1975. Is There a Safer Tea? Morris Arboretum Bull. 26(2): 24-30.

NAS (National Academy of Sciences). 1974. Recommended Dietary Allowances. 8th Revised Edition, NAS, Washington, D.C. 129 pp.

Page, M. and Stearn, W.T. 1979. Culinary Herbs. Wisley Handbook 16, Roayl Horticultural Society, London, 48 pp.

Pauling, L. 1977. Good Nutrition for the Good Life, pp. 77-89 *in* Rodale Press' The Complete Book of Vitamins, Emaus, Pa.

Precheur, R.J. and Garrabrants, N.L. 1983. Weed Control in Dill. The Herb, Spice, and Medicinal Plant Digest 1(2): 3.

Rosengarten, F., Jr. 1973. The Book of Spices. Pyramid Books, New York, 480 pp.

Shimizu, H. 1982. How to Cook with Herbs & Spices, Typescript, National Herb Garden, Washington, 5 pp.

Thomas, C.A. 1953. Herbs and Other Special Crops. USDA Yearbook of Agriculture, pp. 863-868.

Tyler, V.E. 1982. The Honest Herbal, George F. Stickey Co., Philadelphia, 263 pp.

Time-Life, 1977. Herbs. Time-Life Books, Alexandria, 160 pp.

Time-Life, 1982. Beverages. Time-Life Books, Alexandria, 176 pp.

USDA CA 62-24.

USDA. 1963. Composition of Foods. Ag. Handbook No. 8, ARS, USDA, Washington, DC, 190 pp.

Watt, J.M. and Breyer-Brandwijk, M.G. 1962. The Medicinal and Poisonous Plants of Southern and Eastern Africa. 2nd Ed. E. & S. Livingstone, Ltd., Edinburgh and London, 1487 pp.

Wilder, L.B. 1932. The Fragrant Garden. Reprint, Dover, New York, 407 pp.

Williams, L.O. 1960. Drug and Condiment Plants, ARS, USDA, Ag. Handbook 172, USGPO, Washington, 37 pp.

Williams, R.L. 1983. For the All-too-common Cold, We Are Perfect, If Unwilling, Hosts. Smithsonian 14(9): 45-55.

Zanelli, L. 1972. Home Winemaking from A to Z. A.S. Barnes & Co., Cranbury. 135 pp.